Non-Conventional Functional Block Copolymers

ACS SYMPOSIUM SERIES **1066**

Non-Conventional Functional Block Copolymers

Patrick Theato, Editor
Institut für Technische und Makromolekulare Chemie, Universität Hamburg

Andreas F. M. Kilbinger, Editor
Department of Chemistry, University of Fribourg

E. Bryan Coughlin, Editor
Polymer Science and Engineering Department, University of Massachusetts Amherst

Sponsored by the
ACS Division of Polymer Chemistry

American Chemical Society, Washington, DC

Distributed in print by Oxford University Press, Inc.

Library of Congress Cataloging-in-Publication Data

Non-conventional functional block copolymers / Patrick Theato, Andreas F.M. Kilbinger, E. Bryan Coughlin, editor[s] ; sponsored by the ACS Division of Polymer Chemistry.
p. cm. -- (ACS symposium series ; 1066)
Includes bibliographical references and index.
ISBN 978-0-8412-2614-2 (alk. paper)
1. Block copolymers--Congresses. I. Theato, Patrick. II. Kilbinger, Andreas F. M. III. Coughlin, E. Bryan. IV. American Chemical Society. Division of Polymer Chemistry.
QD382.B5N66 2010
547'.7--dc23
2011018396

The paper used in this publication meets the minimum requirements of American National Standard for Information Sciences—Permanence of Paper for Printed Library Materials, ANSI Z39.48n1984.

Distributed in print by Oxford University Press, Inc.

PRINTED IN THE UNITED STATES OF AMERICA

Foreword

The ACS Symposium Series was first published in 1974 to provide a mechanism for publishing symposia quickly in book form. The purpose of the series is to publish timely, comprehensive books developed from the ACS sponsored symposia based on current scientific research. Occasionally, books are developed from symposia sponsored by other organizations when the topic is of keen interest to the chemistry audience.

Before agreeing to publish a book, the proposed table of contents is reviewed for appropriate and comprehensive coverage and for interest to the audience. Some papers may be excluded to better focus the book; others may be added to provide comprehensiveness. When appropriate, overview or introductory chapters are added. Drafts of chapters are peer-reviewed prior to final acceptance or rejection, and manuscripts are prepared in camera-ready format.

As a rule, only original research papers and original review papers are included in the volumes. Verbatim reproductions of previous published papers are not accepted.

ACS Books Department

Contents

Indexes

Preface

The symposium entitled "Non-conventional functional block copolymers" was held at the 239th American Chemical Society Meeting in San Francisco on March 21-25, 2010. We, the organizers, felt that there needed to be a platform for the presentation of polymer science that goes beyond standard (conventional) block copolymers. As "non-conventional" we envisaged (i) stimuli responsive block copolymers, (ii) controlled synthesis of block copolymers, (iii) cyclic block copolymers, (iv) supramolecular assembly to synthesize block copolymers, (v) Janus nanostructures of block copolymers, (vi) new block copolymer thin films, (vii) novel syntheses of block copolymers, (viii) bio-polymer block copolymer conjugates and (ix) organic-inorganic block copolymers, to mention a few of the topics presented at this symposium.

The symposium was intended to be a platform for presenting and exchanging new ideas and concepts in block copolymer science and as such it was a real success. In short, the symposium addressed recent developments in the area of synthesis, characterization and application of block copolymers that are of a non-conventional nature. It addresses novel non-conventional functional block copolymers in all its potential application areas, ranging from bio-sciences to materials sciences. This symposium series collection provides the potential reader the possibility to have a full overview on state of the art applications of block copolymers.

In this symposium book we are delighted to have compiled contributions from selected participants of the symposium, namely Mark Ellsworth, Frederik Wurm, Donghui Zhang, Patrick Theato, Graeme Moad, Elizabeth Gillies, Rajeswari Kasi, John Texter, and Bryan Coughlin.

Mark Ellsworth describes in his chapter how mechanical properties of polymers can be modified by introduction of block and random copolymer architectures. Comparing polyoctadecyl acrylate/polydimethylsiloxane diblock, triblock and random copolymers he was able to show that the random copolymer exhibited the largest strain to break and lowest storage modulus.

Polymers that are only stable in the presence of an end cap and depolymerize in the absence of the latter are termed self-immolative. Elizabeth Gillies describes a general approach to such linear polymers and highlights how amphiphilic block copolymers of self-immolative polymers can be used as slow release containers.

A typical feature of the RAFT process for controlled living radical polymerization is that the RAFT agent has to be tailored depending on monomer reactivity. Graeme Moad describes an alternative approach in his chapter, using RAFT agents that can change reactivity depending on pH. This allows the polymerization of monomers such as vinyl acetate and also styrene or the preparation of block copolymers of the two without changing RAFT agent.

Patrick Theato describes in his chapter the use of orthogonally addressable functionality in diblock copolymers. These copolymers carry alkyne and isothiocyanate groups that can be post-polymerization functionalized by orthogonal, i.e. non-interfering "click-reactions".

By simply attaching Grubbs first generation metathesis catalyst onto amphiphilic block copolymers, Frederik Wurm describes in his chapter how metallated diblock copolymers self-organize into Janus-type micelles in solution.

Using *N*-heterocyclic carbenes to ring open *N*-substituted *N*-carboxyanhydride monomers, rod-coil diblock copolymers were prepared as Donghui Zhang describes in his chapter. Using a combination of *S*-*N*-CHMePh-glycine and *N*-Bu-glycine, block copolymers were formed which consisted of an alpha-helical rod-like segment and a coil-like block. CD-spectroscopy revealed that longer coil-like blocks improved the stability of the helix in the helical block.

John Texter's contribution discusses the incorporation of ionic liquid based monomers as the end-groups of A-B-A triblock copolymers. In comparioins to convetional triblock copolymers these new These triblocks appear more surface active than comparable molecular weight polyethylene oxide-polypropylene oxide-polyethylene oxide triblocks and have lower critical micelle concentrations. They are also anion stimuli responsive and surface activity can be tuned by ion exchange.

The incorporation of side-chain liquid crystals and their shape memory applications are described in Rajeswari Kasi's contribution. Cholestrol based side chains are being investigated for their bioactivity and liquid crystalline properties which can be harnessed for biomedical applications. Thermo-mechanical properties as well as shape memory properties in these polymers are compared, and strategies for outstanding shape memory performance are discussed.

Finally, Bryan Coughlin describes in his chapter how the conventional paradigm of block copolymers most often employ the elements C, H, and perhaps O or N. Non-conventional copolymers can be prepared by expanding the elemental composition of block copolymer by using Boron containing monomers. Initial biological studies have shown the incorporation of the carborane-containing polymers in carcinoma cells. These findings pave the way for future investigations of these polymeric architectures as delivery agents for boron neutron capture therapy of cancer.

We believe that this ACS symposium book provides an excellent snapshot of the exciting research that is going on in the "old" field of block copolymers and that although the basic phase behavior of such systems is well-understood, many more stimulating and exciting aspects of this class of materials has not yet been explored. Finally, we would like to express our sincere gratitude to all presenters and participants of the symposium. It was their devoting contribution that made this symposium a real success. We also wish to recognize the financial contributions from Dow Chemical, ExxonMobil, BMS and PSS. Finally, we would like to thank ACS and in particular ACS books for their support and patience in bringing this book to fruition.

Patrick Theato

Universität Hamburg
Institut für Technische und Makromolekulare Chemie
Bundesstr. 45
20146 Hamburg
Germany

Andreas F. M. Kilbinger

University of Fribourg
Department of Chemistry
Chemin du Musée 9
CH-1700 Fribourg
Switzerland

E. Bryan Coughlin

University of Massachusetts Amherst
Polymer Science and Engineering Department
Amherst, MA 01002
United States

Chapter 1

Tuning the Mechanical Properties of Side Chain Crystallizable Block Copolymers

Dominique M. M. Freckmann,* Anthony (Tony) Idem, and Mark W. Ellsworth

Polymers, Ceramics, and Technical Services Laboratories, Tyco Electronics Corporation, 306 Constitution Dr., Menlo Park, CA 94025
***dominique.freckmann@tycoelectronics.com**

Block and random copolymers based on poly(octadecyl acrylate) (PODA) and polydimethylsiloxane (PDMS) have been synthesized and characterized. With increased PDMS content, the flexibility of the polymers increases, though they continue to exhibit a sharp melting point due to the presence of the PODA segment.

Introduction

Polyacrylates with long aliphatic side chains exhibit a very narrow melt temperature range which can be easily targeted based on the length of the side chain (*1*). However, these highly crystalline materials are brittle, which greatly limits use of these materials. In order to overcome the brittleness of the polyacrylates, copolymers were prepared to introduce rubbery segments having a low glass transition temperature.

The acrylate with the longest side chain which is readily commercially available is octadecyl acrylate which was therefore chosen as monomer for the following copolymer studies. As soft block, polydimethylsiloxane (PDMS) was chosen since a variety of functionalized PDMS are commercially available. Of special interest are the hydroxyethoxypropyl terminated PDMS because they can be converted into halide-capped species which can serve as macroinitiators in ATRP reactions. Additionally, PDMS bearing a methacrylate end group will allow for an easy copolymerization with the acrylate monomer.

To investigate how different amounts of PDMS influence the brittleness of poly(octadecyl acrylate) (PODA, **1**), three PODA/PDMS copolymer structures were synthesized (triblock, diblock, and (random) graft copolymer) and their rheological behavior compared to a commercial sample of pure PODA.

Experimental

Materials

Bis-hydroxyethoxypropyl and mono-methacrylate terminated polydimethylsiloxane (PDMS; molecular weight of ~5,000 g/mol) were purchased from Gelest. Octadecyl acrylate (ODA) was purified by dissolution in hexane, washing with aqueous KOH, drying over $MgSO_4$, passing through neutral alumina, and finally removal of the solvent (*2*). CuBr, $CuBr_2$, 4,4′-dinonyl-2,2′-dipyridyl, 2-bromoisobutyryl bromide, methyl 2-bromopropionate, AIBN (2,2′-azobis(2-methylpropionitrile)), acetone, mesitylene, o-xylene, THF (tetrahydrofuran) were used as received. Pure poly(octadecyl acrylate) (PODA, **1**) was obtained from Landec Corp. (Menlo Park, CA).

Thermal Analysis and Rheology

Melting temperatures (T_m) and Heat of Fusion (H_f) were determined via Differential Scanning Calorimetry (DSC). Gel Permeation Chromatography (GPC) with THF as the mobile phase was utilized to analyze molecular weight and polydispersity of the polymers. Samples (20 x 10 x 2 mm) for Dynamical Mechanical Analysis (DMA) were prepared in a silicone mold by filling in the powders and melting them at 70-80 °C in an oven. Since a significant volume reduction occurs at this point, the process was repeated until the molten polymers fill up the mold completely. Special care had to be taken to avoid the formation of bubbles or cracks upon cooling and releasing from the mold.

To determine thermo-mechanical properties (storage and loss moduli) as functions of temperature, the sample was mounted on a 3-point bending clamp (DMA 2980, TA Instruments) and subjected to a temperature sweep from -150 to 30 °C at 3 °C/minute and oscillated at constant strain amplitude (20 μm) and frequency (1.0 Hz). For measurement of Young's Modulus, Ultimate Strength, Strain-to-Break, and Toughness at room temperature, the sample was held isothermally at 25 °C in a 3-point bending mode and subjected to a linearly increasing stress and the resultant strain was monitored with time.

Synthesis of Triblock Copolymer PODA-*b*-PDMS-*b*-PODA (2)

Following a literature procedure, the esterification reaction of dicarbinol terminated PDMS with α-bromoisobutyryl bromide was carried out to prepare a PDMS-based difunctional macroinitiator (*3*). This macroinitiator was further employed in combination with the iron/phosphine combination developed by Zhu (*4*) to synthesize the triblock PODA-*b*-PDMS-*b*-PODA (**2**) (Figure 1).

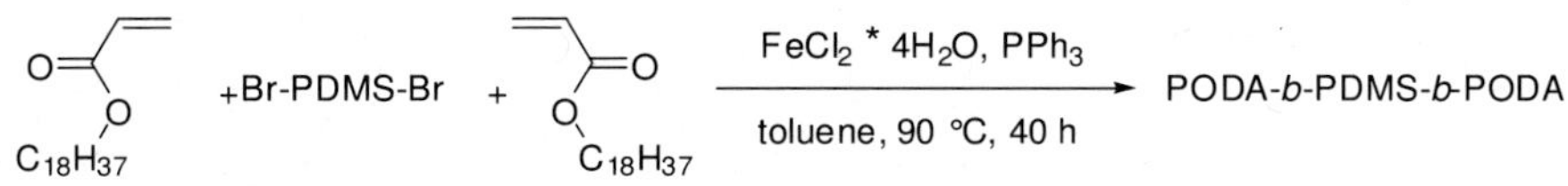

*Figure 1. Synthesis of PODA-b-PDMS-b-PODA (**2**)*

Synthesis of Copolymer PODA-*b*-PDMS (3)

ODA was polymerized using the copper/bipyridine combination developed by Matyjaszewski (*2*). After 40 hours, methacrylate terminated PDMS was added to the reaction mixture and allowed to react for additional 122 hours to give PODA-*b*-PDMS (**3**) (Figure 2).

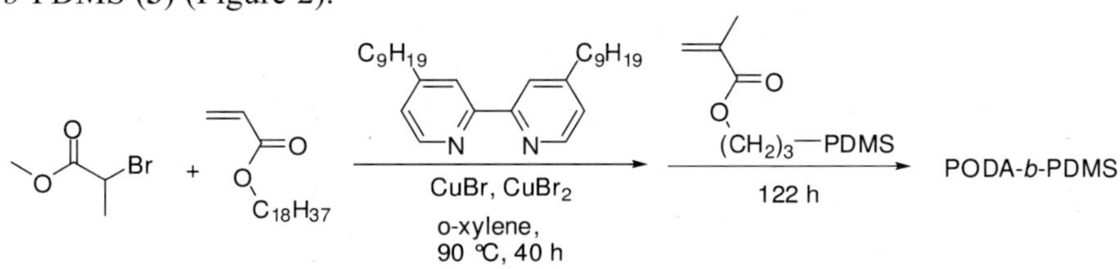

*Figure 2. Synthesis of PODA-b-PDMS (**3**)*

Synthesis of (Random) Graft Copolymer PODA/PDMS (4)

Octadecyl acrylate, mono-methacrylate terminated PDMS, and AIBN were stirred in o-xylene first at room temperature and then heated to 90-110 °C for about 60 minutes until the gas evolution stopped. The clear very viscous solution was diluted with THF and the polymer was isolated by precipitation in acetone (Figure 3).

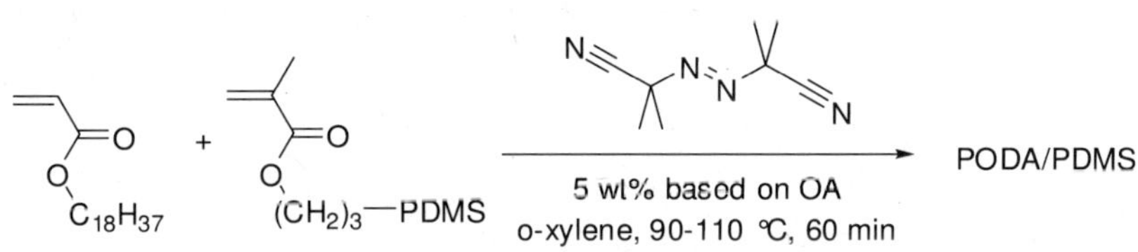

*Figure 3. Synthesis of PODA/PDMS (**4**)*

Results and Discussion

Polymer Synthesis

In order to combine two different desired properties for the new material (flexibility and sharp melting point) it was initially envisaged to synthesize a triblock copolymer. This strategy has proven to bring incompatible polymer systems together causing a microphase segmentation in which each block retains its individual properties (*5*). As polymerization method ATRP was chosen as it allows for a convenient and controlled preparation of well-defined block copolymers with low polydispersity. To obtain triblock copolymers of PODA

Table 1. Reaction conditions, DSC, and GPC data for PODA and PODA/PDMS-copolymers

	Cat	*t; T (h; °C)*	*yield*	T_m (°C)	H_f (J/g)	M_n (g/mol)	M_w (g/mol)	M_w/M_n
PODA-PDMS-PODA (**2**)	[Fe]	40 h; 90°C	46 %	48.6	46.6	15,022	21,466	1.43
PODA-PDMS (**3**)	[Cu]	162 h; 90°C	99 %	48.9	54.1	13,847	19,320	1.40
PODA/PDMS (**4**)	AIBN	1 h; 90°C	81 %	48.0	25.0	19,284	41,413	2.15
PODA (**1**)	n. a.	n. a.	n. a.	50.3	58.7	16,288	54,729	3.36

and PDMS it is the easiest to start from a commercially available PDMS-block and convert it into a halide-capped species which can serve as a macro-initiator. Two different catalyst systems have been used for the ATRP of octadecyl acrylate: a copper/bipyridine combination developed by Matyjaszewski and an iron/phosphine one reported by Zhu (*3*, *4*). Both approaches have shown to work in the present case with slight differences in polymer yields.

In contrast to ATRP, free radical polymerization via a peroxide initiator allows for a fast synthesis of copolymers. AIBN was chosen which upon heating decomposes releasing nitrogen gas and creating two radicals which initiate the free radical polymerization. Because of its uncontrolled fashion, a copolymer is formed with a higher polydispersity than the block copolymers.

Table 1 contains reaction conditions, the DSC, and the GPC data for PODA (**1**) and PODA/PDMS copolymers (**2**), (**3**), (**4**). The polydispersity observed for PODA-*b*-PDMS-*b*-PODA (**2**) is in the same range as reported for PODA-PS-PODA triblock copolymers (PS = polystyrene) (*4*). Though it is larger compared to other typical numbers obtained in ATRP reactions, it is very likely due to the high PDI of the PDMS-macroinitiator (1.70). The PDI of the random **4** is lower than expected, especially compared to the pure PODA which was synthesized by free radical polymerization as well. All polymers exhibit a sharp melting point close to 50 °C but in case of **4** the heat of fusion (H_f) is greatly reduced to around half he value compared to the other three polymers which indicates a much lower degree in crystallinity.

Mechanical Properties at Room Temperature

First, the polymers' response to stress at room temperature was investigated. Figure 4 shows the stress-strain-data for all four polymers and Table 2 lists the values for Young's Modulus, the Ultimate Strength, Strain-to-Break, and Toughness. Though all PDMS-modified PODA copolymers are softer (lower Young's modulus), the commercially available PODA material is still the strongest (highest ultimate strength) and toughest material.

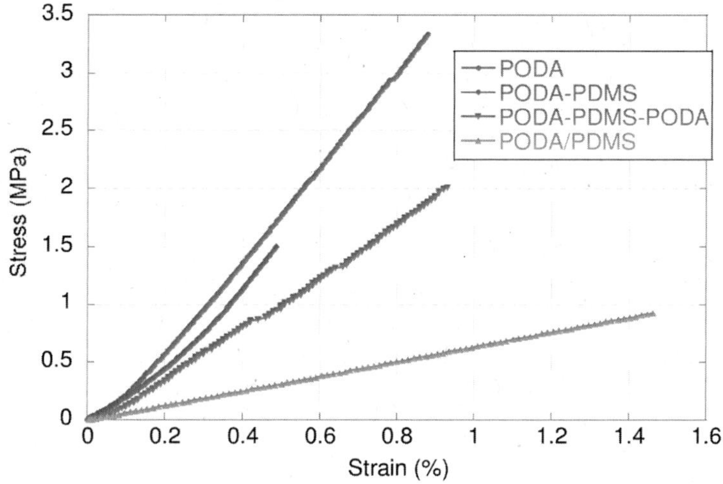

Figure 4. Stress-Strain data at room temperature for PODA and PDMS-copolymers of PODA.

Table 2. Young's Modulus, Ultimate Strength, Strain-to-Break, and Toughness for PODA and PODA/PDMS-copolymers

	Young's Modulus (MPa)	*Ultimate Strength (MPa)*	*Strain-to-Break (Elongation; %)*	*Toughness (N*mm)*
PODA-PDMS-PODA (**2**)	243	2.01	0.9	3.0
PODA-PDMS (**3**)	245	1.50	0.5	1.3
PODA/PDMS (**4**)	64	0.92	1.5	2.4
PODA (**1**)	378	3.34	0.9	5.1

Temperature-Dependent Viscoelastic Properties (E′, E″, Tan δ)

Next, the temperature dependency of the storage modulus E′ (which reflects the amount of energy stored by the material) was examined, which provides indication of the stiffness or the rigidity of the sample. The introduction of PDMS as a co-polymer can lead to a significant decrease of E′ suggesting increased flexibility and suppleness (Fig. 5). The glass transition temperature of the silicone block is easily seen at ~ -120 °C. Another transition is faintly visible around -70 °C.

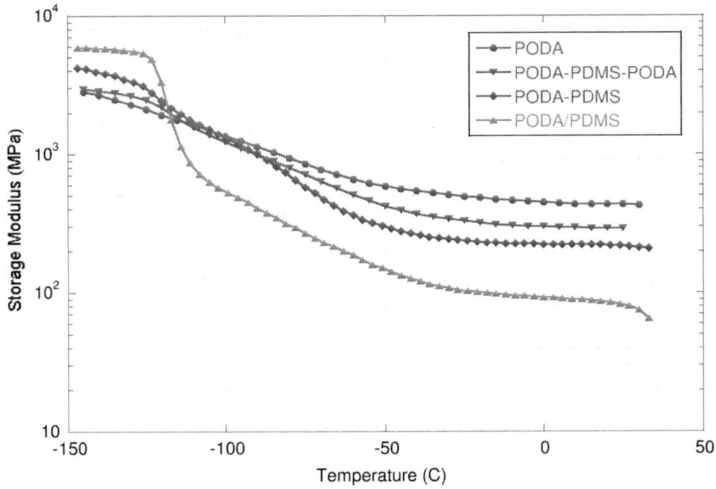

Figure 5. Storage modulus (E') as a function of temperature for PODA and PDMS-copolymers of PODA

During the measurement of the elastic modulus, the loss (or viscous) modulus E″ (which reflects the ability of the material to dissipate energy) is recorded as well. The ratio of E″ to E′ is called tan δ (or damping coefficient) and this is an indicator of how efficiently a material loses energy to molecular rearrangements and internal friction. Figure 6 shows the tan delta curves as a function of temperature and the T_g was determined from the peaks of the tan δ curves. For ease of comparison, all graphs were set relative to the arbitrary value of 0.15 at -150 °C. The T_g for the PDMS in the copolymers is clearly visible around -120 °C which is in accordance with literature values. The second hump between -75 to -60 °C is attributed to PODA related transitions.

Random copolymers would exhibit only one glass transition temperature which would be an intermediate between the T_g's of each individual block. The appearance of two distinct peaks demonstrates the microphase segmentation expected for block and graft copolymers. Though **4** was apparently synthesized in a random fashion via radical polymerization, it exhibits the properties of a graft copolymer. This is due to the large size of the PDMS "monomer" with its polymerizable meth-acrylate functionality at one end. Therefore, the resulting polymer structure should resemble a graft (comb) polymer with a polyacrylate backbone and long PDMS side chains.

In case of PODA-*b*-PDMS-*b*-PODA and PODA-*b*-PDMS, the tan delta peak height of the PDMS –the soft segment– is greatly reduced. This indicates that the soft segment is more restricted in its damping ability suggesting a lower content of the soft part (*6*).

A corresponding correlation between area under the glass transition peaks and content cannot be drawn for the hard part since the curves only relate to the amorphous portion of a polymer. But the large peak, reflecting a high amorphous fraction, seen for PODA/PDMS at -75 °C is in accordance with the lowest crystallinity as measured by DSC (see Table 1).

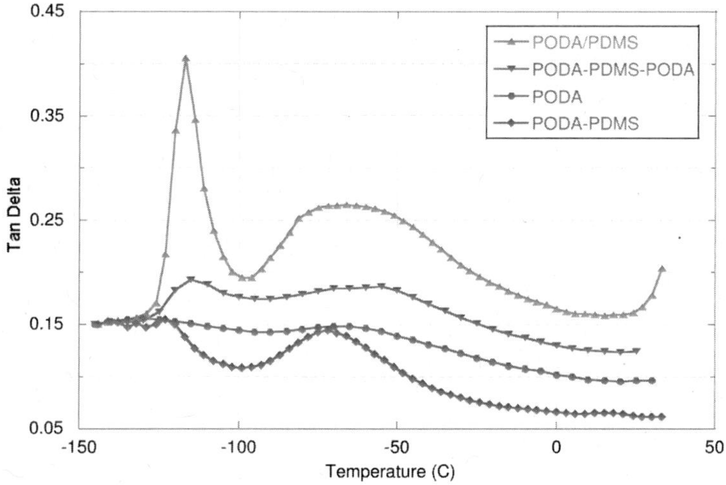

Figure 6. Tan delta curves from DMA temperature sweep for Landec and PDMS-copolymers of PODA.

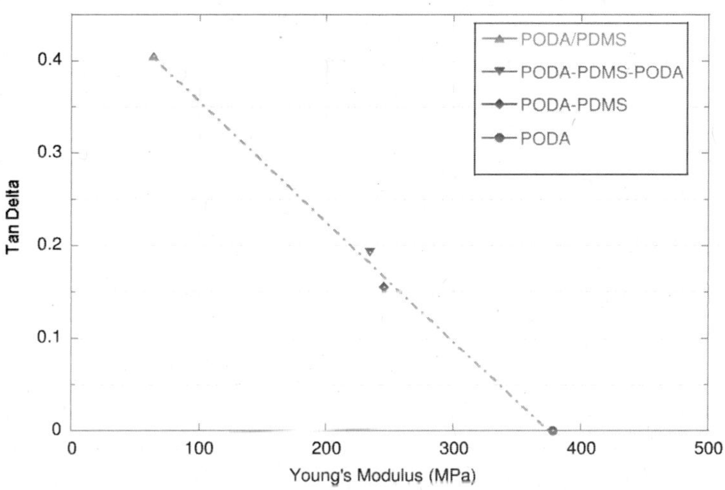

Figure 7. Tan Delta peak values of the PDMS block vs. Young's Modulus at room temperature.

Tan Delta vs. Young's Modulus

As already mentioned, the viscoelastic properties of a material are described by tan delta obtained from the ratio of loss and storage moduli. It is also referred to as the loss tangent and can give information about the damping properties of the material. The greater a tan delta peak the more is the material able to "damp" or dissipate energy. A decrease in its value is related to an increase in stiffness. The intensity of a tan delta peak even at low temperatures can be related to a polymer's impact resistance at room temperature (*7*). Interestingly, a plot of the relative tan delta peak values, associated with the soft PDMS block, against the measure Young's moduli gives a direct relationship (Fig. 7). Therefore, a PDMS block which has a high damping ability is responsible for greater softness at room temperature.

The area under the T_g peaks can be used for a comparison of the relative amounts of amorphous contents. Since the PDMS is completely amorphous, the whole area under the curve represents the PDMS content. Here, PODA/PDMS with its large tan delta peak (high PDMS content) also is the copolymer with the lowest storage modulus, therefore the softest material.

Conclusion

The two PDMS-copolymers of PODA exhibit a lower Young's modulus and therefore are softer than the commercially available PODA but unfortunately, they are not as strong and tough. The PODA/PDMS copolymer, which is the one with the highest silicone content, shows the lowest storage modulus and the largest strain-to-break value. This material was easily synthesized by radical polymerization, hereby avoiding the inconvenient preparation of block copolymers.

Acknowledgments

The authors wish to acknowledge the assistance and support of all of the members of the Tyco Electronics Technology Group, particularly the Materials Characterization Group (Jenny Robison, Gloria Merlino, and Juliana De Guzman) for materials analysis. Stefanie Gravano, Ann Banich, and Allen Nixon are thanked for valuable and fruitful discussions and suggestions.

References

1. Zheng, Q. (J.) Side Chain Mediation of Switchable Properties of Polymers. Presented at Golden Gate Polymer Forum, October 20th, 2004.
2. Qin, S.; Saget, J.; Pyun, J.; Jia, S.; Kowalewski, T.; Matyjaszewski, K. *Macromolecules* **2003**, *36*, 8969.
3. Huan, K.; Bes, L.; Haddleton, D. M.; Khoshdel, E. *J. Polym. Sci., Part A: Polym. Chem.* **2001**, *39*, 1833.
4. Zhu, X.; Gu, Y.; Chen, G.; Cheng, Z.; Lu, J. *J. Appl. Polym. Sci.* **2004**, *93*, 1539.
5. Noshay, A.; McGrath, J. E. *Block Copolymers – Overview and Critical Survey*; Academic Press: New York, NY, 1977.
6. O'Sickey, M. J. Characterization for Structure-Property Relationships of Poly(urethane-urea)s for Fiber Applications. Ph.D. thesis, Virginia Polytechnic Institute and State University, Blacksburg, VA, 2002.
7. Sperling, L. H. *Introduction to Physical Polymer Science*; John Wiley & Sons: New York, NY, 1985.

Chapter 2

Design, Synthesis and Assembly of Self-Immolative Linear Block Copolymers

Matthew A. DeWit,[1] Ali Nazemi,[1] Solmaz Karamdoust,[1] Annelise Beaton,[1] and Elizabeth R. Gillies[1,2,*]

[1]Department of Chemistry, The University of Western Ontario, 1151 Richmond St., London, Canada N6A 5B7
[2]Department of Chemical and Biochemical Engineering, The University of Western Ontario, 1151 Richmond St., London, Canada N6A 5B9
[*]egillie@uwo.ca

In many medical applications it would be useful to trigger the degradation of a polymer under specified physiological conditions, initiating the breakdown of a device, scaffold or the release of encapsulated drugs. Self-immolative linear polymers are a promising approach towards achieving this. They comprise a polymer backbone that is stable in the presence of an end cap, but upon removal of the end cap a cascade of reactions is initiated that results in depolymerization. Described here is a general approach to the development of self-immolative linear polymers and block copolymers. The block copolymers formed self-immolative assemblies in aqueous solution, which could encapsulate and subsequently release their contents in a manner.

Introduction

In recent years, there has been significant interest in the development of biodegradable polymers for diverse biomedical applications ranging from medical devices to tissue engineering scaffolds and drug delivery vehicles. Thus far, significant progress has been made using polymers such as poly(lactic acid) (*1*, *2*), poly(glycolic acid) (*3*, *4*), and polycaprolactone (*5*, *6*). However, these polymers degrade gradually under most physiological and environmental conditions (*7*,

8), and the ability to "turn on" the degradation of these polymers under specific conditions has not yet been demonstrated.

For many applications, it would be desirable to use polymers that can be degraded in a controlled manner in a specified environment or in response to a stimulus such as an enzymatic reaction, or changes in pH or redox potential. This would offer the possibility to use polymer backbones that are stable for extended periods but that will degrade under the desired conditions, resulting in a controlled disintegration of biomedical materials or release of drug molecules from a drug delivery system. Thus far, several polymer backbones containing acetal (*9–11*) or disulfide (*12–14*) linkages have been developed to degrade under mildly acidic or reducing conditions respectively. However, the mechanisms of degradation for these polymers involve random chain scissions throughout the polymer backbone, and many environmentally mediated cleavage events are required to completely degrade the polymer.

End capped cascade degradable or self-immolative polymers are a new and attractive platform for the design of systems that are both degradable and stimuli responsive. As illustrated in Figure 1, these polymers comprise a backbone that is stable when the end cap is intact, but upon removal of the end cap, a functionality is revealed at the polymer terminus that initiates a cascade of intramolecular reactions leading to complete depolymerization from end to end. This concept was initially introduced in dendritic systems, that upon removal of a focal point group were demonstrated to degrade by an intramolecular cascade, releasing multiple molecules such as drugs from the dendrimer periphery (*15–18*). The extension of this concept to linear cascade degradable polymers has the potential to dramatically expand their utility as such materials may be used for the assembly of supramolecular aggregates such as micelles, vesicles, and nanoparticles. In addition, these polymers may impart unique and advantageous properties such as the ability to trigger the cleavage of a single polymer under different conditions by simply tuning the end cap.

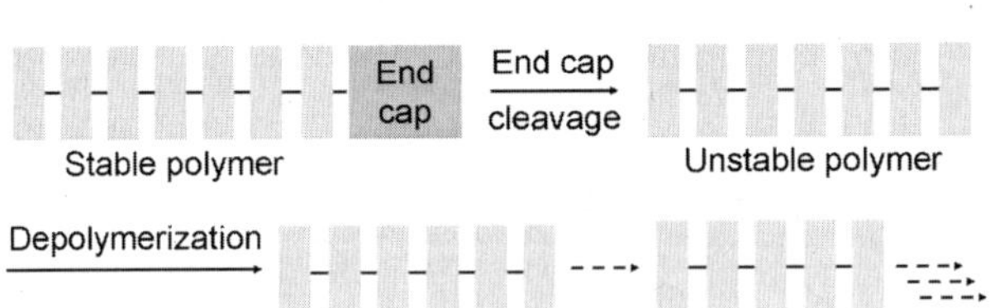

Figure 1. Schematic of a cascade biodegradable polymer. (reproduced from reference (23))

Despite their promise, there are currently very few examples of linear self-immolative polymers. Shabat's group has reported polycarbamates based on 4-aminobenzyl alcohol derivatives (*19–21*). They have shown that they can depolymerize to fluorescent monomers via a series of rapid 1,6-elimination reactions in response to an enzyme mediated end cap cleavage, thus serving as a sensors for enzymatic activity (*19*, *21*). Very recently, Moore and coworkers used a similar polymer backbone for the development of polymer microcapsules that could be release their contents in response to the removal of *t*-butyloxycarbonyl (Boc) or fluorenylmethyloxycarbonyl (Fmoc) triggering groups (*22*). Here we describe a general design strategy for the synthesis of self-immolative linear polymers as well as their degradation by cyclization and elimination reactions (*23*, *24*). The incorporation of these materials into amphiphilic block copolymers and the assembly of these copolymers into nanoparticles under different conditions is also reported.

Results and Discussion

Design

The design of monomers for the synthesis of self-immolative polymers has thus far found inspiration from research on prodrugs, where cleavage of a trigger and subsequent intramolecular reaction of a self-immolative spacer has released the free and active drug from its inactivating group or targeting moiety (*25–28*). In particular, the use of spacers such as hydroxybenzyl alcohol and aminobenzyl alcohol derivatives in the development of self-immolative polymers is attractive as they undergo rapid 1,4- or 1,6-elimination reactions when the free aromatic alcohol or amine functionalities are revealed (Figure 2a) (*29*). In addition, these functionalities can be readily masked when the molecules are activated, thus providing relatively stable polymerization monomers and synthetic intermediates. Another potential class of monomers for self-immolative polymers is those capable of undergoing intramolecular cyclization reactions (*26*). The incorporation of these monomers is attractive as they can potentially be used to modulate the depolymerization rate. Furthermore, it has been suggested that the quinone methide intermediates involved in the above mentioned elimination reactions can potentially lead to toxicity (*30*, *31*), so it would be desirable to develop new backbones that do not involve hydroxybenzyl alcohol or aminobenzyl alcohol. One example of a cyclization spacer is a carbamate derivative of *N,N'*-dimethylethylenediamine, which spontaneously cyclizes to form *N,N'*-dimethylimidazolidinone (Figure 2b) (*32*). The cyclizations of 2-mercaptoethanol derivatives to the corresponding cyclic thiocarbonate are also of interest (Figure 2c) (*33*, *34*).

The development of monomers capable of undergoing polymerization to form cascade degradable polymers requires careful design. In particular, in the preparation of polymers designed to degrade by cyclization mechanisms, cyclization of the activated monomer (Figure 2d) must be avoided. We have found that the synthesis and polymerization of activated dimers is an effective approach, as the activated leaving group is distant from the nulceophilic moiety

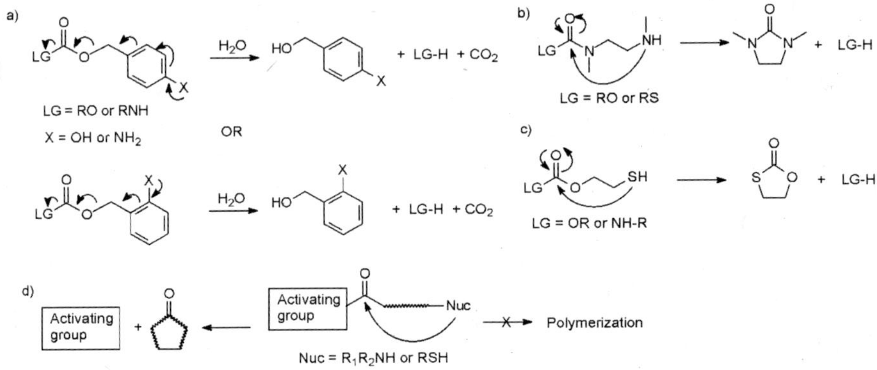

Figure 2. Self-immolative spacers: a) 1,4- and 1,6-eliminations; b) cyclization of N,N'-dimethylethylenediamine derivatives; c) cyclization of 2-mercaptoethanol derivatives; d) competition between polymerization and intramolecular cyclization.

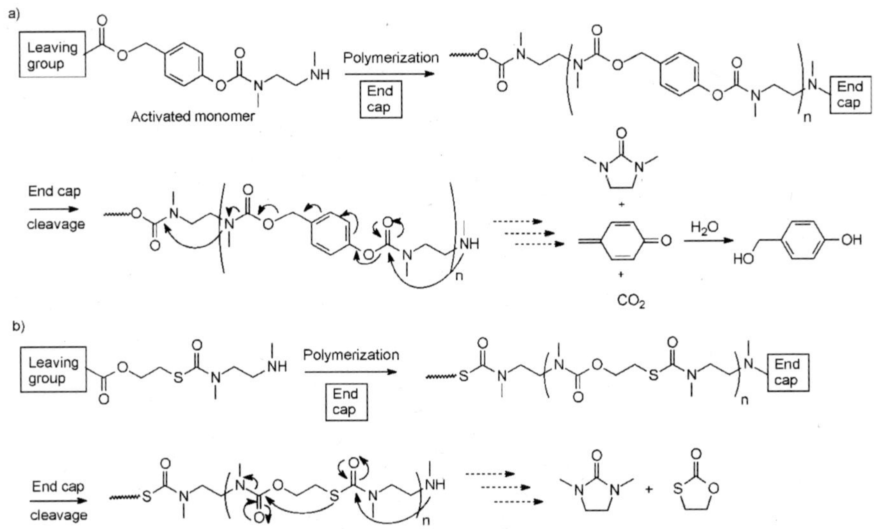

Figure 3. Design of self-immolative polymers degrading by: a) alternating 1,6-elimination and cycliation reactions; b) cyclization reactions.

such that the resulting ring size is not particularly favourable for cyclization. This allows polymerization to be a highly competitive reaction at high concentrations.

Based on these principles, two monomers were designed. The first (Figure 3a) incorporated *N,N'*-dimethylethylenediamine and 4-hydroxybenzyl alcohol (*23*). Upon removal of the end cap from the resulting polymer terminus, cyclization of the diamine should occur, releasing *N,N'*-dimethylimidazolidinone and revealing the phenol. The phenol should then undergo a 1,6-elimination, ultimately releasing 4-hydroxybenzyl alcohol, CO_2, and another amine terminus, from which the cascade can continue until depolymerization is complete. The second (Figure 3b) was based on 2-mercaptoethanol and *N,N'*-dimethylethylenediamine (*24*).

In this case, removal of an end cap from the polymer would lead to alternating cyclization reactions with the release of *N,N'*-dimethylimidazolidinone and 1,3-oxathiolan-2-one. Use of a polymeric end cap in either case, would lead to a block copolymer.

Polymer and Block Copolymer Synthesis

Polymerization of monomer **1** with 5 mol% of the Boc protected end cap **2** in the presence of 4-(dimethylamino)pyridine (DMAP) and *N,N*-diisopropylethylamine (DIPEA) provided polymer **3** (Figure 4a) (*23*). A monomer to end cap ratio of approximately 16:1 was determined by ^{1}H NMR analysis. Although a Boc group cannot be readily cleaved under any known physiological conditions, it can be easily cleaved with TFA under non-aqueous conditions, allowing the depolymerization process to be studied independently of the end cap cleavage. On the other hand, polymerization of **1** with the poly(ethylene oxide) (PEO) end cap **4** provided the block copolymer **5** (Figure 4b). The ratio of monomer to end cap in the resulting polymer was approximately 15:1 based on ^{1}H NMR spectrscopy. An ester linkage that could readily be hydrolyzed under neutral physiological conditions was selected for the conjugation of the PEO block to the terminus of the cascade degradable block, but this concept could readily be extended to pH or redox sensitive linkages in order to initiate the degradation cascade under a wide range of conditions.

For the preparation of a polymer degrading entirely by cyclization reactions, the polymerization of monomer **6** with 2 mol% the disulfide end cap **7** was performed to provide polymer **8** (Figure 4c) (*24*). Using ^{1}H NMR spectroscopy, a ratio of end cap to monomer of approximately 35:1. As the disulfide can be cleaved under mildly reducing conditions, this polymer was expected to be a redox responsive self-immolative polymer that could be cleaved selectively within cells or hypoxic tumor tissue where there are elevated levels of biological reducing agents such as glutathione (*14*, *35*, *36*). In initial work, the resulting polymers were purified by preparation size exclusion chromatography (SEC). However more recent efforts have demonstrated that they can also be purified by dialysis, a much less labour-intensive procedure. NMR analysis of the products as well as the dialysate did not reveal any evidence of monomer cyclization for any of the above polymers, indicating that the intermolecular polymerization reaction successfully competed with this potential side reaction.

Polymer Degradation

To study the degradation of polymer **3**, the polymer was treated with TFA/CH_2Cl_2 to remove the Boc end cap, and then the polymer was incubated in 0.1 M phosphate buffer:acetone (3:2) at 37 °C (*23*). The degree of degradation was quantified by ^{1}H NMR spectroscopy. Peaks corresponding to 4-hydroxybenzyl alcohol and *N,N'*-dimethylimidazolidinone emerged as the degradation progressed. This supports that the degradation occurred by the proposed depolymerization mechanism as a degradation mechanism based mainly on random chain scission of the carbamate linkages in the polymer backbone

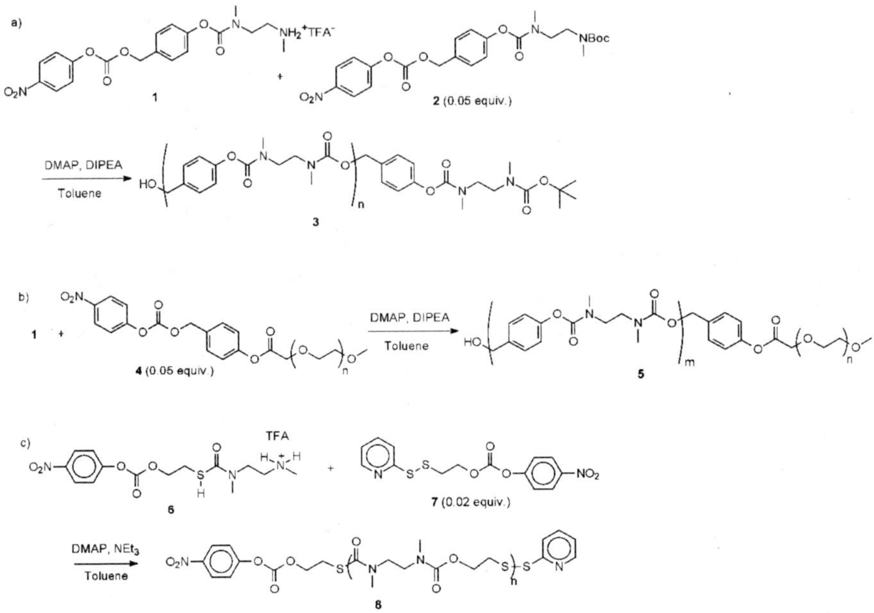

Figure 4. Syntheses of a) a polymer based on alternating benzyl alcohol and N,N'-dimethylethylenediamine; b) a block copolymer based on PEO and the same polymer as above; c) a polymer based on N,N'-dimethylethylenediamine and 2-mercaptoethanol.

would generate primarily *N,N'*-dimethylethylenediamine (not observed) rather than the cyclic urea. The degradation was 50% complete in less than one day, and complete degradation was observed after 4-5 days (Figure 5a). The fact that the depolymerization reached completion suggests that there was no significant amount of nondegradable cyclic polymers. A control sample of polymer **3** in which the Boc end cap was left intact exhibited only trace levels of degradation after 4 days.

The depolymerization of polymer **8** was studied by first dissolving the polymer in pH 7.4 phosphate buffered D_2O:acetone-d_6 (3:2) at 37 °C and then cleaving the disulfide end cap by the addition of dithiothreitol (DTT) (*24*). The reducing conditions were maintained by periodic additions of DTT. The degradation was monitored by ^1H NMR spectroscopy and characteristic peaks appeared corresponding to *N,N'*-dimethylimidazolidinone and 1,3-oxathiolan-2-one. Again, the presence of these products is a strong indicator that the degradation proceeds by the proposed cascade of cyclization reactions as random chain scissions would lead to *N,N'*-dimethylethylenediamine and 2-mercaptoethanol, products that were not detected in the NMR spectra. In addition, a control sample incubated under the same conditions except in the absence of DTT did not reveal the appearance of any degradation products, thus indicating the end cap cleavage was required to initiate the degradation. As shown in Figure 5b, the degradation reached 80% completion after 10-14 days. However, no significant further degradation was observed, even after 30 days suggesting that the residual

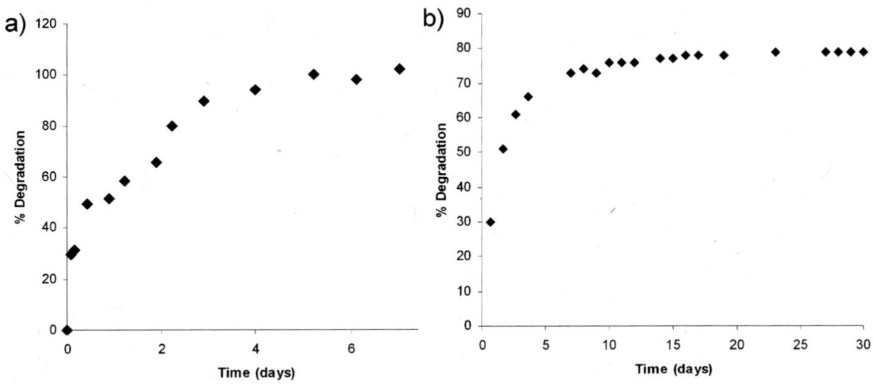

*Figure 5. Degradation kinetics for polymers a) **3** and b) **8**.*

polymer was cyclic species. The presence of cyclic species was also supported by SEC analysis of the polymer and its degradation products.

Block Copolymer Assembly

Solubility tests on polymer **3** revealed that it was insoluble in pure water or buffered aqueous solutions and soluble in a wide range of organic solvents such as chloroform, dichloromethane, THF, and DMF. Based on this, as well as the known water solubility of PEO, it was anticipated that the block copolymer **5** might be amphiphilic and self-assemble in aqueous solution. Indeed, when **5** was added to pure water or 0.1 M phosphate buffer, the polymer did not immediately dissolve, but with sonication it readily dispersed. Transmission electron microscopy (TEM) performed using phosphotungstic acid as a stain revealed the presence of nearly spherical aggregates with a relatively broad distribution of diameters ranging from less than 100 nm to a few hundred nm (Figure 6a). Assemblies in water were also detected by dynamic light scattering (DLS) (Figure 7a). The extreme broadness and very low intensities of the peaks corresponding to the hydrophobic polycarbamate block of **5** in its ^{1}H NMR spectrum in D_2O was additional evidence of the polymer's assembly.

The assemblies induced by dissolving polymer **5** in THF and then gradually adding water and dialyzing away the THF were also investigated. By TEM (Figure 6b), nanoparticles with diameters of approximately 50 nm were observed. This size, also supported by DLS measurements (Figure 7b) is within the range expected for micelles, indicating that the larger nanoparticles observed above may have been kinetically trapped aggregates. However, these smaller micelles did exhibit some tendency to aggregate, as observed by both TEM and DLS.

Degradation of Block Copolymer Assemblies

The depolymerization of the above polymer assemblies was monitored by ^{1}H NMR spectroscopy in 0.1 M phosphate buffered D_2O at 37 °C (*23*). As described above for polymer **3**, peaks corresponding to 4-hydroxybenzyl alcohol

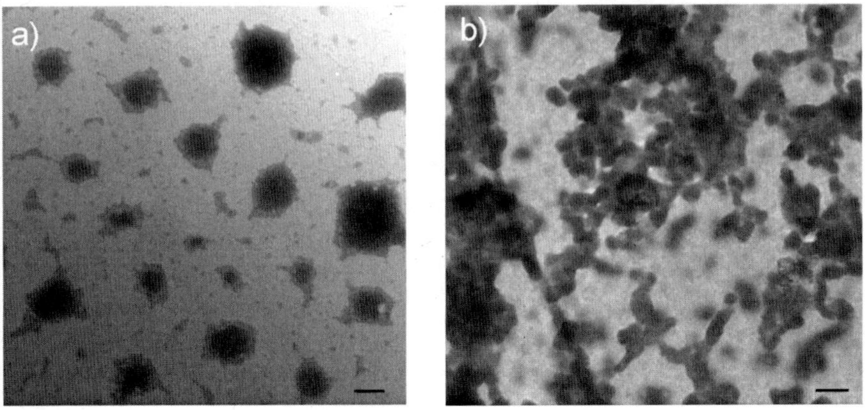

*Figure 6. Transmission electron micrscopy images of a) assemblies formed by the direct sonication of **5** in water; b) assemblies formed by the dissolution of **5** in THF followed by addition of water. (scale bar = 100 nm)*

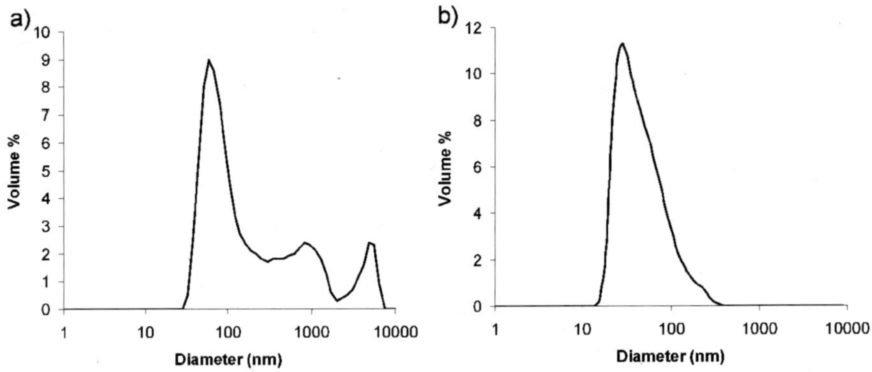

*Figure 7. Dynamic light scattering analyses of a) assemblies formed by the direct sonication of **5** in water; b) assemblies formed by the dissolution of **5** in THF followed by addition of water.*

and *N,N'*-dimethylimidazolidinone emerged as the degradation progressed and the kinetic data are shown in Figure 8a. Even taking into account the time required for the removal of the PEO end cap via ester hydrolysis which had a half life of 15 hours (*23*), the depolymerization rate is clearly slower than that observed for polymer **3**. This can likely be attributed to the formation of the above described assemblies. While the depolymerization of **3** was carried out in a water:acetone mixture in which the polymer was fully dissolved, the depolymerization of **5** occurred at the hydrophobic cores of the assemblies. A slowing of reactions with polar transition states in the hydrophobic cores of micelles has been previously reported (*37*, *38*) and is consistent with the observation that the diamine cyclization was quite dependent on the polarity of the solvent. These results indicate that the rate of the degradation cascade can be modulated not only by incorporating new monomers with different depolymerization rates, but also

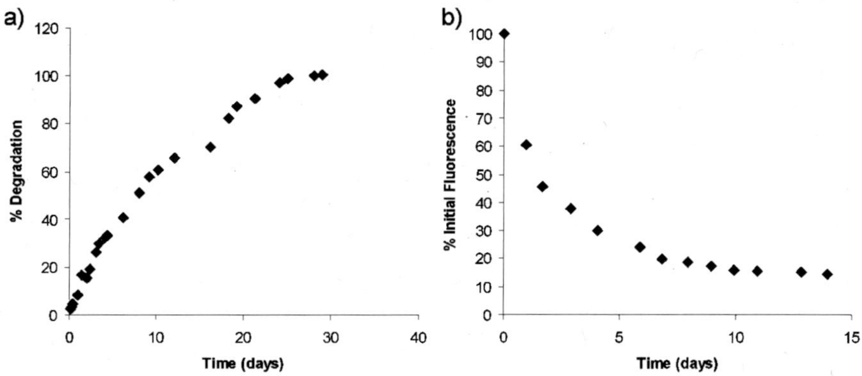

Figure 8. a) Degradation kinetics for assemblies formed by copolymer **5***; b) fluorescence decrease of Nile Red corresponding to its release from assemblies of copolymer* **5***.*

by tuning the hydrophobicities of the materials and controlling their assembly into nanoaggregates.

Controlled Release Properties of the Block Copolymer Assemblies

The degradation kinetics of the assemblies formed by polymer **5** suggest that they may be useful for slow, controlled release applications. To evaluate this possibility, the dye Nile Red was encapsulated as a model hydrophobic drug and its release was studied (*23*). Its fluorescence is negligible in aqueous solutions but increases substantially in the hydrophobic compartments of polymer assemblies (*39*, *40*). The encapsulation was performed by sonicating the polymer assemblies in pH 7.4 phosphate buffer in the presence of insoluble Nile Red. Some Nile Red was taken up as evidenced by a significant increase in the fluorescence of the solution in comparison with the negligible fluorescence of Nile Red which was sonicated in phosphate buffer alone. The loaded nanoparticles were then incubated in this buffer at 37 °C, allowing depolymerization to occur. Over a time period of two weeks, the fluorescence of the dye relative to a control solution of Nile Red, decreased to less than 15% of its initial fluorescence, consistent with its release from the nanoparticles (Figure 8b). The time scale of this release was similar to that of the polymer degradation. This experiment therefore suggests that nanoparticles comprising cascade degradable polymers can provide well controlled release properties that depend on the rate of depolymerization.

Conclusions

A general approach to the development of self-immolative linear polymers was developed based on the synthesis and polymerization of activated dimers. The use of a polymeric end cap was demonstrated to provide block copolymers. Two different polymer backbones were synthesized with degradation mechanisms based on either alternating cyclization and elimination cascades or entirely cyclization reactions. It is noteworthy that the depolymerization kinetics for both of these polymers were found to be significantly slower than those of the previously reported backbone based entirely on 1,6-elimination spacers (*19*, *20*). Furthermore the rate was modulated by the assembly of the block copolymers into nanoparticles. These rates open up many new potential applications such as drug carrier systems or biomedical devices where a prolonged degradation or release of molecules is required. The incorporation of additional monomers and end caps, as well as the use of these interesting materials for biomedical applications is currently under investigation and will be reported in due course.

Experimental

General Procedures and Materials

All reagents were purchased from commercial suppliers and used without further purification unless otherwise noted. Anhydrous toluene was obtained from a solvent purification system. NEt_3 was distilled from CaH_2. NMR spectra were obtained at 400 MHz using a Varian Mercury or Varian Inova spectrometer. Chemical shifts are reported in ppm and are calibrated against residual solvent signal of $CDCl_3$ (δ 7.26, 77.36). Coupling constants are expressed in Hertz (Hz). SEC (DMF) was carried out at a flow rate of 1 mL/min in DMF with 10 mM LiBr and 1% (v/v) NEt_3 at 85 °C using a Waters 2695 separations module equipped with a Waters 2414 differential refractometer and two PLgel 5 μm mixed-D (300 mm × 7.5 mm) columns from Polymer Laboratories. SEC (THF) was performed using a Waters 515 pump equipped with an Optilab Rex refractive index detector from Wyatt Technology and two Resipore (300 mm × 7.5 mm) columns from Polymer Laboratories. The calibrations were performed using polystyrene standards. Dynamic light scattering analyses were performed using a Zetasizer Nano ZS from Malvern Instruments.

General Polymerization Procedure

The monomer (1.0 equiv) and end cap (0.05 or 0.02 equiv) were dissolved in anhydrous toluene (monomer concentartion = 0.2 M) and freshly distilled NEt3 or DIPEA (5.0 equiv) and DMAP (0.1 equiv) were added. The reaction was stirred for 24 h at room temperature. The solution was then diluted with CH_2Cl_2, washed once with 1 M HCl, then twice with 10% aqueous Na_2CO_3. The organic layer was dried over $MgSO_4$ and the solvent was removed in vacuo to provide to provide the crude product. This material was then purified by dialysis in DMF (Spectrum Laboratories Spectra/Por, 12-14 kDa MW cutoff).

Polymer **3**. ^1H NMR ($CDCl_3$): δ 7.35 (br s, 2H), 7.12 – 7.00 (m, 2H), 5.15 – 5.03 (m, 2H), 3.66 – 3.41 (m, 4H), 3.15 – 2.83 (m, 6H), 1.50 – 1.40 (m, 0.5H). SEC (DMF): M_n = 17 000, M_w = 26 900, PDI = 1.6.

Polymer **5**. ^1H NMR ($CDCl_3$): δ 7.34 (br s, 2H), 7.12 – 7.01 (m, 2H), 5.16 – 5.05 (m, 2H), 4.42 (s, 0.1H), 3.85 – 3.41 (m, 71H), 3.38 (s, 0.3H), 3.15 – 2.89 (m, 6H). SEC (DMF): M_n = 28 700, M_w = 35 000, PDI = 1.2.

Polymer **8** (no extraction performed during workup). ^{1}H NMR (400 MHz, $CDCl_3$): δ 8.48 (d, *J* = 4.7), 8.28 (d, *J* = 9.0) 7.74 – 7.61 (m, 2H), 7.41 (*d*, J = 9.0), 7.10 (br s, 1H), 4.39 – 4.28 (m, 2H), 4.27 – 4.07 (m, 76H), 3.61 – 3.34 (m, 155H), 3.23 – 3.10 (m, 77H), 3.02 (br S, 114H), 2.98 – 2.89 (m, 120H). SEC (THF): M_n = 1800 g/mol, M_w = 2950 g/mol, PDI = 1.6.

Block Copolymer Assembly

Method A: The block copolymer (3 mg) was dissolved in CH_2Cl_2 and the solution was passed through a 0.2 μm syringe filter. The solvent was removed in vacuo, 1 mL of H_2O was filtered into the flask, and the mixture was sonicated for 30 seconds. *Method B:* The block copolymer (3 mg) was dissolved in 300 μL of filtered THF. 1 mL of deionized water was added dropwise over 10 minutes to the THF solution with vigorous stirring (1100 rpm). In order to remove THF, the resulting suspension was dialyzed against deionized water for 48 hours (Spectrum Laboratories Spectra/Por, 12-14 kDa MW cutoff).

Transmission Electron Microscopy (TEM)

The suspension prepared by Method A or B (20 μL) was deposited onto a carbon formvar grid and was left for 4 minutes. The excess solution was then blotted away using a strip of filter paper. 20 μL of 2% aqueous phosphotungstic acid (PTA) solution was added on the grid and left to stand for 30 seconds. The excess PTA solution was blotted again with a strip of filter paper and the sample was left to dry in air for 30 minutes prior to imaging. Images were obtained using a Phillips CM10 microscope operating at 80kV with a 40 μm aperture.

Encapsulation and Release of Nile Red (*23*)

The block copolymer **5** (2 mg) and Nile Red (0.2 mg) were dissolved in 1 mL of CH_2Cl_2 and mixed thoroughly. The solvent was then removed. 2 mL of pH 7.4 0.1 M phosphate buffer solution was added, and the mixture was sonicated for two hours. The fluorescence was measured immediately after sonication, after which the sample was incubated at 37°C. A second sample, containing only Nile Red, was treated in exactly the same manner as the polymer solution to obtain a background fluorescence value for Nile Red in phosphate buffer. A control sample of Nile Red was also prepared by dissolving 0.1 mg in 1 mL of THF and diluting by 1500x to obtain a fluorescence value similar to that of the polymer. The fluorescence of the degrading sample relative to the control sample was measured daily to determine the percentage of the initial fluorescence while correcting for instrumental fluctuations. The fluorescence spectra were obtained on a QM-4 SE spectrometer from Photon Technology International (PTI), equipped with double excitation and emission monochromators. An excitation wavelength of 550 nm was used for Nile Red and the emission spectra were recorded from 565 and 700 nm.

Acknowledgments

The authors thank the Natural Sciences and Engineering Research Council of Canada, the Canada Research Chairs Program, the Canada Foundation for Innovation, and the Ontario Research Fund for support of the research.

References

1. Langer, R.; Vacanti, J. P. *Science* **1993**, *260*, 920–926.
2. Vert, M.; Schwach, G.; Engel, R.; Coudane, J. *J. Controlled Release* **1998**, *30*, 85–92.
3. Bala, I.; Hariharan, S.; Kumar, M. N. *Crit. Rev. Ther. Drug Carrier Syst.* **2004**, *21*, 387–422.
4. Astete, C. E.; Sabliov, C. M. *J. Biomater. Sci., Polym. Ed.* **2006**, *17*, 247–289.
5. Bezwada, R. S.; Jamiolkowski, D. D.; Lee, I.; Vishvaroop, A.; Persivale, J.; Treka-Benthin, S.; Erneta, M.; Suryadevara, J.; Yang, A.; Lui, S. *Biomaterials* **1995**, *16*, 1141–1148.
6. Pektok, E.; Nottelet, B.; Tiller, J. C.; Gurny, R.; Kalagos, A.; Moeller, M.; Walpoth, B. H. *Circulation* **2008**, *118*, 2563–2570.
7. Vert, M. *Biomacromolecules* **2005**, *6*, 538–546.
8. Sun, Y.; Watts, D. C.; John, J. R.; Shukla, A. J. *Am. Pharm. Rev.* **2001**, *8*, 10–16.
9. Hefferman, M. J.; Murthy, N. *Bioconjugate Chem.* **2005**, *16*, 1340–1342.
10. Jain, R.; Standley, S. M.; Frechet, J. M. J. *Macromolecules* **2007**, *40*, 452–457.
11. Murthy, N.; Thng, Y. X.; Schuck, S.; Xu, M.; Frechet, J. M. J. *J. Am. Chem. Soc.* **2002**, *124*, 12398–12399.
12. Emiltiri, E.; Ranucci, E.; Ferruti, P. *J. Polym. Sci., Part A: Polym. Chem.* **2005**, *43*, 1404–1416.
13. Christensen, L. V.; Chang, C.-W.; Kim, W.-J.; Kim, S. W.; Zhong, Z.; Lin, C. C.; Engbersen, J. F. J.; Feijen, J. *Bioconjugate Chem.* **2006**, *17*, 1233–1240.
14. Meng, F.; Hennick, W. E.; Zhong, Z. *Biomaterials* **2009**, *30*, 2180–2198.
15. Amir, R. J.; Pessah, N.; Shamis, M.; Shabat, D. *Angew. Chem., Int. Ed.* **2003**, *42*, 4494–4499.
16. de Groot, F. M. H.; Albrecht, C.; Koekkoek, R.; Beusker, P. H.; Scheeren, H. W. *Angew. Chem., Int. Ed.* **2003**, *42*, 4490–4494.
17. Szalai, M. L.; Kevwitch, R. M.; McGrath, D. V. *J. Am. Chem. Soc.* **2003**, *125*, 15688–15689.
18. Avital-Shmilovici, M.; Shabat, D. *Soft Matter* **2010**, *6*, 1073–1080.
19. Sagi, A.; Weinstain, R.; Karton, N.; Shabat, D. *J. Am. Chem. Soc.* **2008**, *130*, 5434–5435.
20. Weinstain, R.; Sagi, A.; Karton, N.; Shabat, D. *Chem. Eur. J.* **2008**, *14*, 6857–6861.
21. Weinstain, R.; Baran, P. S.; Shabat, D. *Bioconjugate Chem.* **2009**, *20*, 1783–1791.

22. Esser-Kahn, A. P.; Sottos, N. R.; White, S. R.; Moore, J. S. *J. Am. Chem. Soc.* **2010**, *132*, 10266–10268.
23. DeWit, M. A.; Gillies, E. R. *J. Am. Chem. Soc.* **2009**, *131*, 18327–18334.
24. DeWit, M. A.; Beaton, A.; Gillies, E. R. *J. Polym. Sci., Part A: Polym. Chem.* **2010**, *48*, 3977–3985.
25. Bundgaard, H. *Adv. Drug Delivery Rev.* **1989**, *3*, 39–65.
26. Shan, D.; Nicolaou, M. G.; Borchardt, R. T.; Wang, B. *J. Pharm. Sci.* **1997**, *86*, 765–767.
27. Tranoy-Opalinski, I.; Fernandes, A.; Thomas, M.; Gesson, J.-P.; Papot, S. *Anti-Cancer Agents Med. Chem.* **2008**, *8*, 618–637.
28. Filupa, D.; Zhao, H. *Adv. Drug Delivery Rev.* **2008**, *60*, 29–49.
29. Wakselman, M. *Nouv. J. Chim.* **1983**, *7*, 439–447.
30. Monks, T. J.; Jones, D. C. *Curr. Drug Metab.* **2002**, *3*, 425–438.
31. Guyton, K. Z.; Thompson, J. A.; Kensler, T. W. *Chem. Res. Toxicol.* **1993**, *6*, 731–738.
32. Saari, W. S.; Schwering, J. E.; Lyle, P. A.; Smith, S. J.; Engelhardt, E. L. *J. Med. Chem.* **1989**, *33*, 97–101.
33. Satyam, A. *Bioorg. Med. Chem. Lett.* **2008**, *18*, 3196–3199.
34. Meyer, Y.; Richard, J. A.; Massonneau, M.; Renard, P. Y.; Romieu, A. *Org. Lett.* **2008**, *10*, 1517–1520.
35. Saito, G.; Swanson, J. A.; Lee, K. D. *Adv. Drug Delivery Rev.* **2003**, *55*, 199–215.
36. Kuppusamy, P.; Li, H.; Ilangovan, G.; Cardounel, A. J.; Zweier, J. L.; Yamada, K.; Krishna, M. C.; Mitchell, J. B. *Cancer Res.* **2002**, *62*, 307–312.
37. Gillies, E. R.; Jonsson, T. B.; Fréchet, J. M. J. *J. Am. Chem. Soc.* **2004**, *126*, 11936–11943.
38. Chan, Y.; Wong, T.; Byrne, F.; Kavallaris, M.; Bulmus, V. *Biomacromolecules* **2008**, *9*, 1826–1836.
39. Krishna, M. M. G. *J. Phys. Chem. A* **1999**, *103*, 3589–3595.
40. Gillies, E. R.; Fréchet, J. M. J. *Chem. Commun.* **2003**, 1640–1641.

Chapter 3

Orthogonally Reactive Diblock Copolymers Utilizing Alkyne and Isothiocyanate Groups

Peter J. Roth[1,2] and Patrick Theato*,[1,3,4]

[1]Institute of Organic Chemistry, University of Mainz, Duesbergweg 10-14, 55099 Mainz, Germany
[2]*current address*: Centre for Advanced Macromolecular Design, School of Chemical Science and Engineering, University of New South Wales, UNSW Sydney NSW 2052, Australia
[3]WCU program of Chemical Convergence for Energy & Environment (C2E2), School of Chemical and Biological Engineering, College of Engineering, Seoul National University (SNU), Seoul, Korea
[4]Department of Materials Science and Engineering, The University of Sheffield, Mappin Street, Sheffield, S1 3JD, UK
***theato@uni-mainz.de**

The monomer 4-isothiocyanato styrene (ITS) was synthesized and in combination with trimethyl((4-vinylphenyl)ethynyl)-silane (TMVES) polymerized under controlled radical polymerization conditions resulting in orthogonally reactive block copolymers. These were employed in the preparation of covalently bound layer-by-layer thin films allowing for sequential control of each layer deposited. A precise control of the internal film at the nanometer scale was achieved.

Introduction

Block copolymers are a fundamentally important class of polymers due to their intrinsic properties, which arise from the different chemical nature of each individual block (*1*). Block copolymers are receiving great attention, as they are known for their microphase separation behavior that can yield unique periodic nanostructures in bulk or in many self-assembled structures in solution. Diblock copolymers are generally synthesized by living or controlled polymerization techniques of two monomers. Thus the functionality of the resulting polymers is defined by the monomers used. To further enhance the chemical functionalities of block copolymers, the introduction of defined functional groups is an important topic of current polymer synthesis. Utilizing monomers with reactive side groups for modern controlled polymerization techniques has provided new opportunities in the preparation of well-defined functional polymers and block copolymers. Several review articles have recently summarized the chemical possibilities in using reactive monomers for controlled polymerization techniques, which highlight the range of applicable reactive chemical groups (*2–5*).

Synthesizing polymers with orthogonally reactive groups provides the possibility to modify polymers with more than one functional group independently, yielding new highly functional block copolymers with interesting properties. So far, different examples of functionalization of diblock copolymers have been described in the literature, for example (see Figure 1) combining alkynyl groups (for dipolar cycloadditions) and alcohols (for esterification reactions) (*6*), alkynyl groups and activated esters (for amidations) (*7*), or activated esters and disulfides (for thiol exchange reactions) (*8*). Also orthogonal reactions using aldehydes and azides (*9*), lactone groups and allylethers (*10*), as well as hydroxyl groups with common orthogonal protecting groups (*11*) have been described. An alternative route employs selective hydrogen-bonding interactions to non-covalently bind various functional groups (*12*). Theato et al. recently demonstrated a sequential conversion of block copolymers featuring activated ester side groups on the basis of pentafluorophenyl esters (*13*).

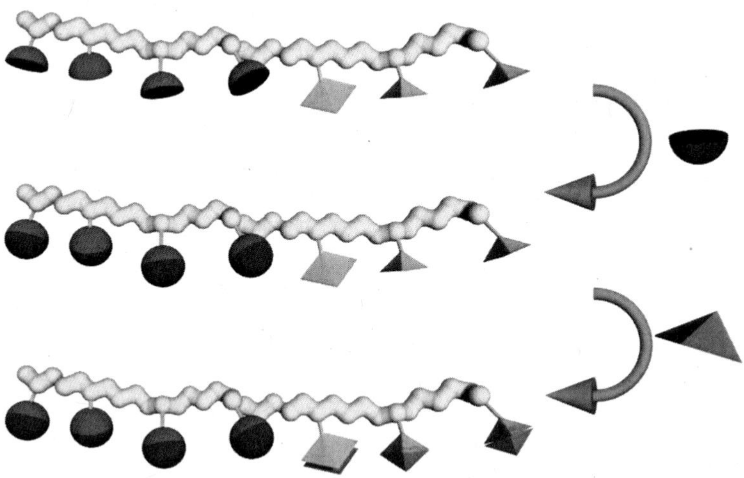

Figure 1. General schematic for the synthesis of functional diblock copolymers utilizing an orthogonally reactive precursor diblock copolymer.

In the present work, we describe a new reactive monomer, 4-isothiocyanato styrene, and its polymerization. To the best of our knowledge, isothiocyanate containing polymers have not been prepared by controlled polymerization techniques. The preparation of diblock copolymers of 4-isothiocyanato styrene in combination with trimethyl((4-vinylphenyl)ethynyl)silane will be conducted resulting in orthogonally reactive diblock copolymers. Its conversion via two orthogonal "click reactions", copper catalyzed azide / alkyne cycloaddition (CuAAC) after deprotection of the alkyne group, and amine with isothiocyanate, will be examined and the application in multilayer formation will be elaborated.

Results and Discussion

Within the present study it is anticipated to combine two orthogonally reactive groups in a diblock copolymer. We chose the isothiocyanate group because it selectively reacts with amines and is known to be stable against water and alcohols. The alkyne group is known for a 1,3-dipolar cycloaddition with azides under copper(I) catalysis.

Monomer Synthesis

At first, the two monomers 4-isothiocyanato styrene (ITS) and trimethyl((4-vinylphenyl)ethynyl)silane (TMVES) had to be synthesized, as shown in Figure 2. ITS could be prepared starting from 2-(*p*-aminophenyl)-ethanol in two steps. First, 2-(*p*-aminophenyl)-ethanol was dehydrated by heating it to 200°C over KOH. 4-Aminostyrene could be distilled off under vacuum. Purification by a second vacuum distillation resulted in pure 4-aminostyrene in 43% yield. The isothiocyanate was then formed by reaction of 4-aminostyrene with carbon disulfide in the presence of *N,N'*-dicyclohexylcarbodiimide. Distillation resulted in pure isothiocyanato styrene in 54% yield (Figure 2A).

Trimethyl((4-vinylphenyl)ethynyl)silane (TMVES) was synthesized according to the literature (*6*). Briefly, 4-bromo styrene was used for a Sonogashira coupling with ethynyltrimethylsilane using bis(triphenyl phosphine) palladium(II) chloride and copper(I) iodide as the catalytic system. TMVES was obtained after vacuum distillation as a liquid in 59% yield (Figure 2B).

Polymerization of TMVES

The styrene derivative carrying the protected triple bond, TMVES, was polymerized either by free radical polymerization, or through a nitroxide mediated polymerization (NMP) (see Figure 3). The free radical polymerization was carried out with AIBN as initiator at 90°C. Both bulk polymerization and solvent polymerization using dioxane as solvent were employed. High molecular weights ranging up to 131 kg/mol could be obtained in high yields in polymerization times as short as 70 min through free radical bulk polymerization. Using dioxane as solvent drastically decreased the polymerization rate and the yield. TMVES could also successfully be polymerized by NMP, giving polymers with a low molecular weight distribution and functionalized chain ends. For example, TMVES was heated with 1-(1-phenylethoxy)-2,2,6,6-tetramethylpiperidine to 120°C for 3.5 hours, yielding poly(TMVES) (PTMVES) of 8200 g/mol and a dispersity index of 1.16. Similarly, a polymer of 14000 g/mol and a dispersity index of 1.29 was prepared. In order to show the functionality of the chain end group, the 8.2 kg/mol PTMVES was employed as a macro unimer to mediate a polymerization of styrene. This served as a preliminary experiment to determine the optimal procedure to produce the desired polymer consisting of a TMVES block and an ITS block. The polymerization of styrene mediated by 8.2 kg/mol TEMPO-terminated PTMVES was carried out at 120°C in bulk styrene for 3 hours. A product with a molecular weight of 75.2 kg/mol was obtained suggesting a successful grafting of a polystyrene block onto PTMVES. However, the product had a dispersity index of 1.53, suggesting the occurrence of unwanted side reactions. It was therefore next investigated whether a block formation in the opposite order would lead to a better product. Consequently, styrene was polymerized under NMP conditions as mentioned above yielding polystyrene (PS) of 23.4 kg/mol and a PDI of 1.22. This PS was then in turn used to mediate a polymerization of TMVES giving the expected diblock copolymer PS-*b*-PTMVES with a molecular weight of 70.6 kg/mol and a dispersity of 1.41. This being a reasonably low value for the given high molecular weight, it was concluded that TMVES should better be used as the second monomer for the synthesis of diblock copolymers.

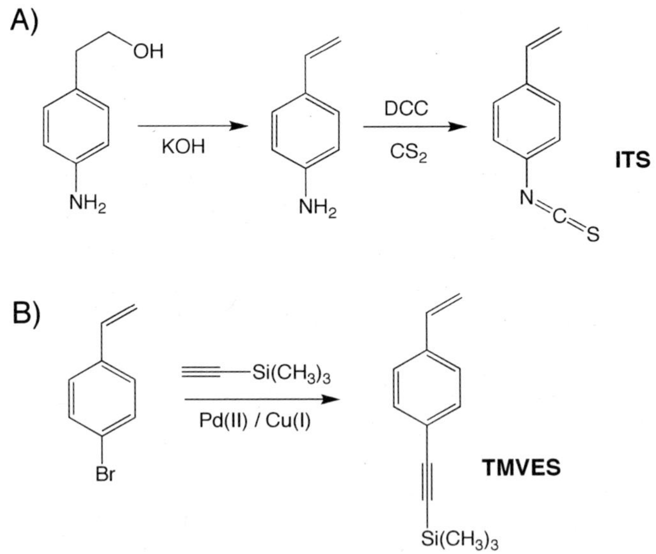

Figure 2. Synthesis of (a) isothiocyanate styrene (ITS) and (b) trimethyl((4-vinylphenyl)ethynyl)silane (TMVES).

Polymerization of ITS

In the next step, the optimal conditions to polymerize the new, orthogonally functional ITS monomer were investigated (see Figure 3). ITS was first polymerized under free radical polymerization conditions in dioxane at 90°C using AIBN as the radical initiator. Due to a poor solubility of poly(ITS) (PITS) in dioxane, the polymerization caused some precipitation. Nonetheless, PITS was obtained essentially in quantitative yield by precipitation into methanol, with a molecular weight of 61.5 kg/mol and a dispersity of of 2.11, very close to the expected theoretical value of 2. Alternatively, a free radical polymerization in THF at 60°C gave comparable results after 2 days. On the other hand, a bulk polymerization of ITS at 90°C resulted in a crosslinked gel after 30 min. It was assumed that ITS easily decomposed at elevated temperatures. Accordingly, a thermogravimetric analysis was carried out (Figure 4). It was found, that the thermal decomposition did indeed start above 100°C, with a 5% mass loss occurring until the sample had been heated to 200°C. A more rapid mass loss with temperature started above 350°C. This thermal sensitivity thus limits the use of polymerization techniques. The ^{1}H-NMR spectrum of PITS prepared in dioxane or THF both showed broad polystyrenic peaks in the range of 7.3 - 6.3 ppm, with the polymer backbone appearing at 2.0 - 1.0 ppm. Furthermore, the FT-IR spectrum proved that the isothiocyanate group was still present due to the very intensive band at 2100 cm^{-1}, which is very indicative for the cumulative –N=C=S bond. This demonstrated the stability of the –N=C=S group against alcohols, very much in contrast to isocyanates (–N=C=O), which are known to readily react with alcohols resulting in the respective carbamates.

The solubility of PITS was then determined. PITS was readily soluble in THF, dichloromethane, DMF and *N*-Methyl-2-pyrrolidone. In contrast, it was insoluble in water, methanol, ethanol, DMSO, ethyl acetate, acetonitrile, chloroform, toluene, and hexane.

In order to (co-)polymerize ITS via NMP, several test polymerizations were carried out to assess how the thermal instability of the monomer would influence polymerization kinetics at 120°C. After 3 h of heating bulk ITS with the TEMPO unimer, PITS of 19.5 kg/mol was obtained in high yield. However, the dispersity index was 1.61. Although considerably lower than a free radical polymerization, this distribution is rather high for a controlled radical polymerization and suggests the occurrence of unwanted side reactions, probably due to thermal decomposition of the isothiocyanate groups. Other conditions were thus investigated. It was found, that side reactions producing high PDIs could be limited, if an NMP was not carried out in bulk but in the presence of a solvent, such as freshly distilled 1,4-dioxane. Additionally, reversible addition fragmentation chain transfer (RAFT) polymerization in the presence of benzyl dithiobenzoate was probed as a polymerization technique. As RAFT can be performed at considerably lower temperatures, a lower PDI was expected, while still maintaining a controlled polymerization. ITS could successfully be polymerized by RAFT at 60°C in THF, giving PITS of 8.5 kg/mol and a s dispersity of 1.37 in a 60% yield (Figure 3).

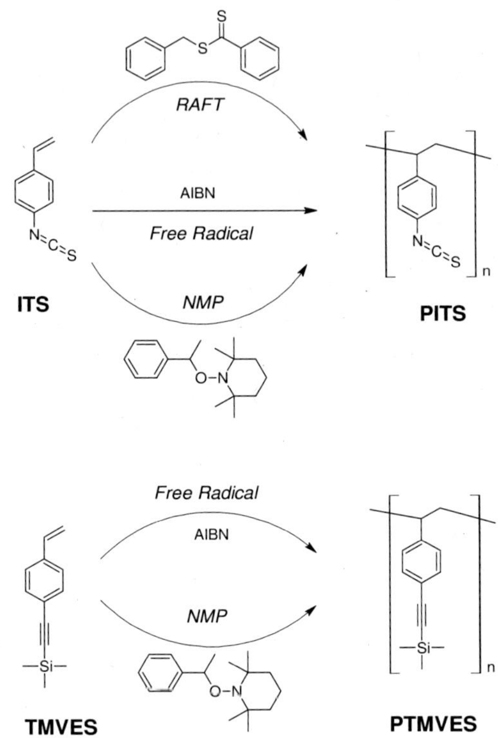

Figure 3. Polymerization techniques employed to polymerize isothiocyanato styrene (ITS) and trimethyl((4-vinylphenyl)ethynyl)silane (TMVES).

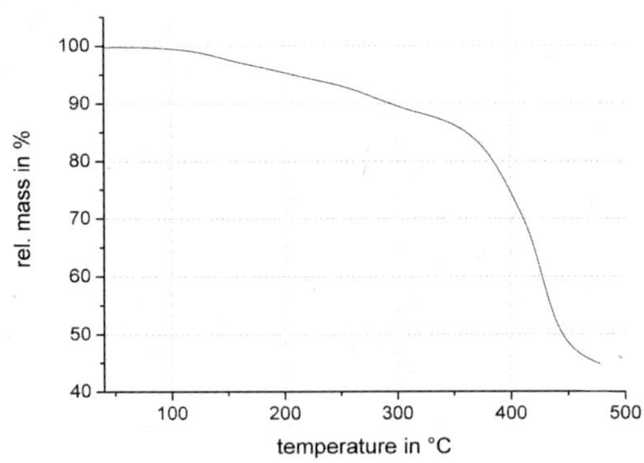

Figure 4. Thermogravimetric analysis of the monomer ITS, showing its decomposition above 100°C.

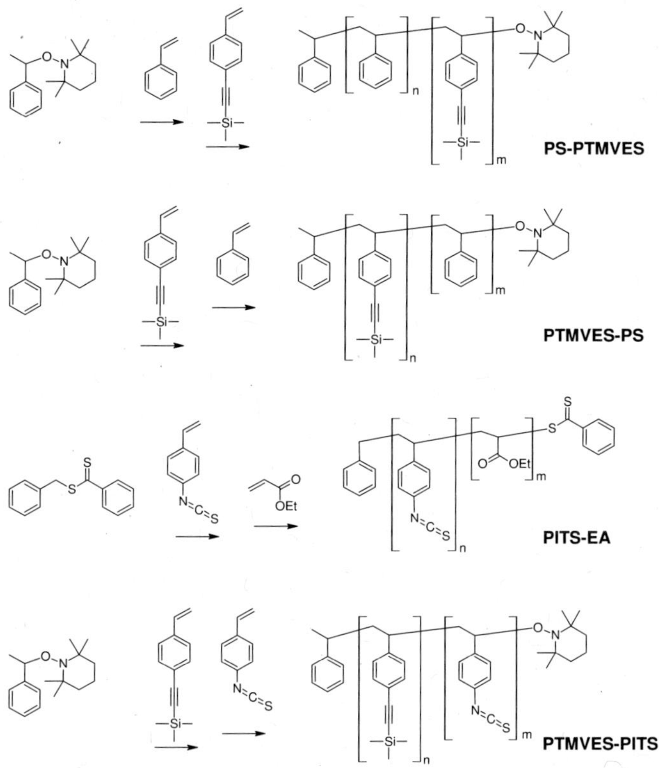

Figure 5. Synthesis various diblock copolymers discussed in the text, including the orthogonally reactive diblock copolymer poly(trimethyl((4-vinylphenyl)ethynyl)silane)-block-poly(isothiocyanato styrene).

Synthesis of Diblock Copolymers

Nitroxide mediated polymerization was investigated as method to prepare diblock copolymers consisting of an ITS and a TMVES block (see Figure 5). Although preliminary experiments with styrene and TMVES had shown that a better polymerization control could be achieved by polymerizing TMVES as a second block mediated by a nitroxide terminated polystyrene, TMVES was used as a first block for the synthesis of the desired double functional diblock copolymer. This way, heating of the sensitive monomer ITS would be limited only to its own polymerization. A PTMVES block with a molecular weight of 14 kg/mol and a dispersity of 1.29 was used to mediate the polymerization of ITS at 120°C for 3 hours. The reaction was carried out with dioxane as solvent so as to reduce crosslinking side reactions of the PITS block. A polymer of a molecular weight of 23.7 kg/mol and a PDI of 1.61 was obtained, suggesting a composition of 14 g/mol PTMVES (70 repeating units) and 9.5 kg/mol PITS (60 repeating units), if the molecular weights based on polystyrene calibration are taken as absolute molecular weights. NMR integration of the trimethylsilyl groups compared to the aromatic signals indicated a 48 % fraction of TMVES, showing a nearly 50:50 composition of the two orthogonally reactive groups.

After the successful synthesis of the desired difunctional reactive diblock copolymer, the separate addressability of each block was probed (see Figure 6). Standard procedures for the functionalization could be applied, however, a careful selection of the order of functionalization had to be considered. As such, it seemed that conversion with an amine should be conducted first, followed by the deprotection and conversion of the alkyne. A detailed study on the selective functionalization is beyond the scope of the present publication and will be published elsewhere.

Layer-by-Layer (LbL) Assembly

Layer-by-layer (LbL) assembly (*14*, *15*) is a versatile fabrication method for the preparation of thin films that find various applications such as chemical sensors (*16*), drug delivery vehicles (*17–19*), and solid electrolyte membranes (*20*, *21*). For the LbL deposition mainly noncovalent intermolecular interactions, such as Coloumbic interactions (*22*) and hydrogen-bonding interactions (*23*, *24*), have been employed. Even though the concept has already been described in the patent of Decher et al., only very recently have LbL films on the basis of covalent bonds between adsorbing chains been realized (*25–29*). Covalently bonded LbL films show a higher mechanical and chemical stability than noncovalently bonded LbL films. Only a selected number of studies have been reported on covalently bonded LbL systems, but to the best of our knowledge, no orthogonal bonding motive to potentially auto-control the thin film deposition has been investigated so far.

For that purpose, the orthogonally reactive block copolymer polymer PTMVES-*b*-PITS seemed to be an ideal candidate. A glass slide was dip-coated into a solution of poly(ethylene imine) (PEI) in ethanol, providing an amine functionalized surface through stable physisorption of PEI to the glass. Subsequently, the amine modified glass slide was immersed into a solution of the difunctional diblockcopolymer PTMVES-*b*-PITS in THF (see Figure 7). Reaction of the polymer isothiocyanate groups with the amines on the surface created a covalent attachment of the polymer to the glass slide, preserving the trimethylsilyl protected functional groups. A UV-vis absorbance spectrum of the glass slide was recorded. After all surface functionalization steps the glass slide was thoroughly washed to remove any non-covalently bound materials. In the next step, the acetylene moieties were deprotected immersing the slide into a tetrabutylammonium fluoride solution, following a literature procedure (*30*). Another UV-vis spectrum was measured. This spectrum was subtracted from the first UV-vis measurement and also from the following measurements made, i.e., it was used as a reference. The spectra are shown in Figure 8. The difference of the spectrum before acetylene deprotection and after (black curve in Figure 5B (color online)) clearly shows the absorbance of the trimethylsilyl protecting group (260 nm) (*31*) that is removed through the deprotection step, thus freeing the reactive triple bonds. The absorbance of the deprotected diblock covalently attached to the surface was set to zero (red curve (color online)). Next, the free triple bonds on the surface were reacted with a complementary reactive

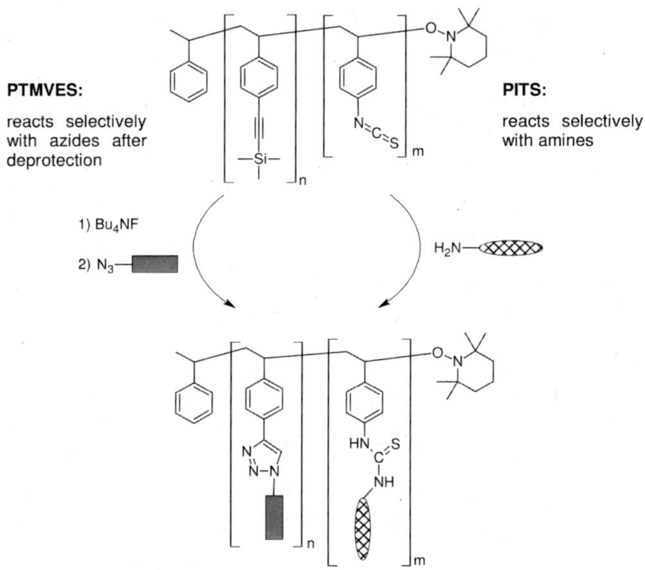

Figure 6. General reaction scheme for the selective addressability of the polymer PTMVES-b-PITS by two orthoganol click reactions.

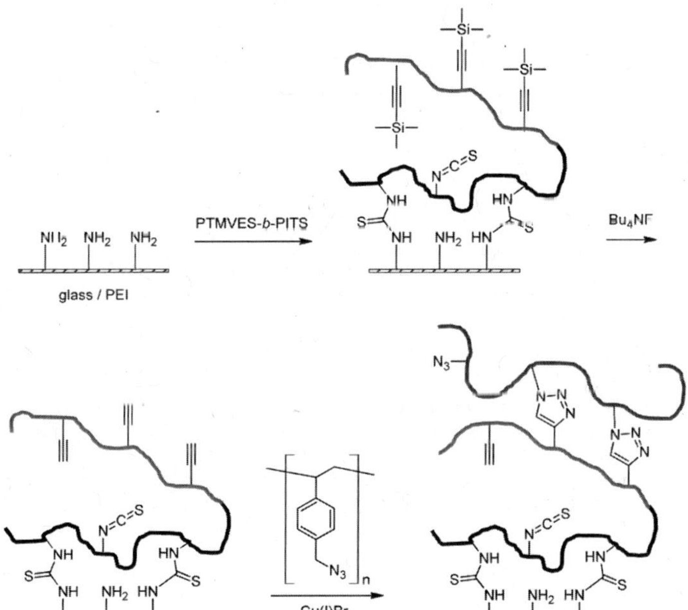

Figure 7. Schematic for the covalent attachment of reactive polymers to a glass surface through dip coating.

polymer, poly(*p*-azidomethyl styrene) (PAzMS). This allowed introducing a different reactive functionality onto the surface, as not all azide groups are converted into triazoles for steric reasons. A successful covalent deposition

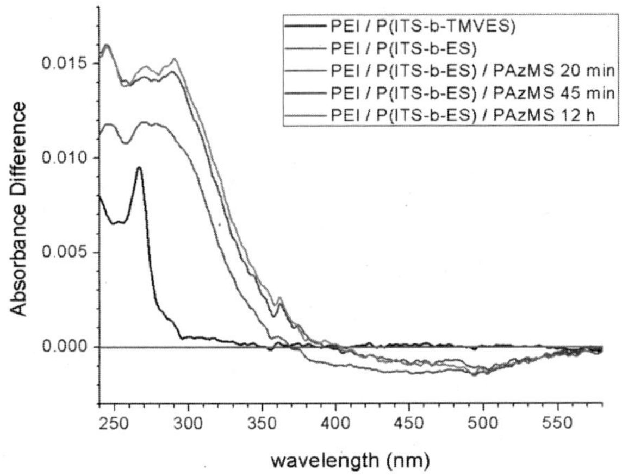

Figure 8. UV-vis absorbance curves of a glass slide functionalized according to Figure 7 and as discussed in the text. The absorbance of the deprotected polymer poly(isothiocyanato styrol-b-ethinylstyrol),P(ITS-b-ES)) was subtracted from all measurements as a reference for better clarity.

of PAzMS thus provides the possibility for multi-layered systems based on orthogonal functionalities and featuring remaining functionalities inside the film, to be addressed by low molecular weight materials diffusing into the film. The deposition of PAzMS was monitored by UV-vis absorbance. First, the glass slide with the bifunctional polymer exposing triple bonds was immersed into a THF solution of PAzMS containing a trace of Cu(I)Br as catalyst for 20 min. A UV-vis spectrum after subtraction of the reference spectrum showed an increase of absorbance below 350 nm, typical for the aromatic rings in polystyrene and derivatives (*32*) (blue curve in Figure 8 (color online)). Evidently, additional polymeric material had been covalently attached to the reactive surface through a CuAAC clicking reaction. In order to determine the completeness of the surface clicking reaction, the glass slide was again immersed into the solution containing Cu(I)Br and the azide modified polystyrene derivative for another 25 minutes, allowing for a total reaction time of 45 minutes. After thoroughly rinsing the slide, another UV-vis measurement (purple curve in Figure 8 (color online)) was recorded. Another increase in aromatic absorbance was observed, suggesting that the reaction had not been completed after 20 minutes. The glass slide was again returned into the CuAAC clicking solution and left to sit at room temperature overnight. Interesting, the UV-vis spectrum measured after thorough rinsing of the slide now (dark yellow curve in Figure 8 (color online)) did not differ from the one measured after 45 minutes. This data shows that the CuAAC reaction could successfully be used to covalently deposit a new reactive layer within 45 minutes. Thus, both orthogonally reactive groups of the polymer could independently be addressed at room temperature, showing the click-like ease and versatility of the two reactive groups incorporated into a single diblock copolymer.

Experimental Part

Materials

All chemicals were commercially available and used as received unless otherwise stated. 2,2′-Azoisobutyronitrile (AIBN) was recrystallized from diethylether. All solvents were purified by common procedures. 1,4 Dioxane was freshly distilled from sodium.

Instrumentation

^{1}H and ^{13}C NMR spectra were obtained on a Bruker AC 300 MHz FT-NMR spectrometer. Gel permeation chromatography (GPC) was used to determine molecular weights and polydispersity indices (PDI), M_w/M_n, of polymeric samples with respect to polystyrene standards. Therefore, a GPC set-up was used consisting of the following components: a Jasco PU-1580 pump, a Jasco AS-1555 autosampler, MZ-Gel-SDplus columns (102, 104 and 106 Å2), a Jasco RI-1530 refractive index detector, and a Jasco UV-1575 UV/vis detector. Infrared (IR) spectra were measured on a Bruker Vector 22 FT-IR spectrometer with ATR unit.

Trimethyl((4-vinylphenyl)ethynyl)silane (TMVES)

16.8 g (91.3 mmol) 4-bromo styrene, 15.6 mL (109 mmol) ethynyltrimethylsilane, 2.1 g bis(triphenyl-phosphine) palladium(II) chloride and 7.1 mg copper(I) iodide were dissolved in 370 mL pyrrolidine. The solution was stirred at 50°C for 8 hours and afterwards the organic solution was washed with water and then concentrated in vacuum. TMVES was then obtained as a pure liquid after two consequtive vacuum distillations. Yield 10.8 g (53.5 mmol, 59%). ^{1}H-NMR, 300 MHz, $CDCl_3$ δ (ppm) = 7.42 (d, 2 H, Ar-H, J = 8 Hz), 7.33 (d, 2 H, Ar-H, J = 8 Hz), 6.68 (dd, 1 H, Ar-CH=CH_2, J_1 = 18 Hz, J_2 = 11 Hz), 5.45 (d, 1 H, Ar-CH=CH_2, J = 18 Hz), 5.28 (d, 1 H, Ar-CH-CH , J = 11 Hz), 0.26 (s, 9 H, Si-(CH)). ^{13}C-NMR δ (ppm) = 137.6 (Ar-<u>C</u>H-CH_2), 136.2(vinyl-ipso), 132.1 (2 C, vinyl-meta), 126.0 (2 C, vinyl-ortho), 122.4 (vinyl-para), 114.8 (Ar-CH-<u>C</u>H_2), 105.1(Ar-<u>C</u>CSi), 94.8 (Ar-C<u>C</u>Si), -0.01 (Si(<u>C</u>H_3)$_3$).

4-Amino Styrene

30 g 2-(*p*-aminophenyl)-ethanol were heated over 90 g KOH to 200°C and the raw product (22 g) was slowly distilled off under light vacuum. Redistillation in high vacuum yielded pure 4-amino styrene in 11.3 g (43%) yield. n_D = 1.6227; ^{1}H-NMR, 300 MHz, $CDCl_3$ δ (ppm) = 7.35 (d, 2 H, aromatic H), 6.65 (m, 3H, aromatic H + CH), 5.58 (d, 1H, CH_2), 5.08 (d, 1H, CH_2), 3.65 (s, 2 H, NH_2).

4-Isothiocyanato Styrene (ITS)

11.3 g 4-amino styrene was added under ice cooling to 19.5 g *N,N′*-dicyclohexylcarbodiimide, 38 g carbon disulfide and 19 mL pyridine and the mixture was then stirred for 10 hours at room temperature. After filtration of the urea, the solution was extracted with diethylether and the organic phase was distilled under vacuum yielding 8.3 g ITS as a liquid. n_D = 1.678; ^{1}H-NMR, 300 MHz, $CDCl_3$ δ (ppm) = 7.36 (m, 2 H, aromatic H), 7.15 (m, 2 H, aromatic H), 6.66 (dd, 1 H, CH), 5.73 (d, 1 H, CH_2), 5.29 (d, 1H, CH_2). FT-IR: 2100 cm^{-1} (-NCS).

General Procedure for Free Radical Polymerization

An example is given: 500 mg ITS and 1 mg AIBN were dissolved in 5 mL dioxane and heated to 90°C for 8 h. PITS was then precipitated into methanol and obtained in quantitative yield. ^{1}H-NMR, 300 MHz, $CDCl_3$ δ (ppm) = 7.3 - 6.3 (bd, 4 H, aromatic H), 2.0 – 1.0 (b, 3 H, polymer backbone). FT-IR: 2100 cm^{-1} (-NCS). GPC M_n = 61500 g/mol, M_w/M_n = 2.11.

General Procedure for Nitroxide Mediated Polymerization (NMP)

An example is given: A mixture of 0.95 mL TMVES and 7.3 mg TEMPO unimer (1-(1-phenylethoxy)-2,2,6,6-tetramethylpiperidine) were degassed in a Schlenk flask and then stirred for 2 hours at 120°C. The obtained polymer was then dissolved in a small amount of THF and then precipitated into methanol. Yield: 390 mg (46%). GPC M_n = 14000 g/mol, M_w/M_n = 1.29. ^{1}H-NMR, 300 MHz, $CDCl_3$ δ (ppm) = 7.00-7.25 (m, 2 H, Ar-H), 6.10- 6.50 (m, 2 H, Ar-H), 1.05-1.90 (m, 3 H, CH-CH_2), 0.24 (bs, 9 H, $Si(CH_3)_3$).

General Procedure for the Synthesis of Diblock Copolymers

The synthesis followed the general NMP procedure given above, where the TEMPO unimer is replaced with the nitroxide terminated first block.

Poly[styrene-*block*-trimethyl((4-vinylphenyl)ethynyl)silane] (**PS-*b*-PTM-VES**). Made from polystyrene (GPC M_n = 23400 g/mol, M_w/M_n = 1.22), GPC (diblock) M_n = 70600 g/mol, M_w/M_n = 1.41.

Poly[trimethyl((4-vinylphenyl)ethynyl)silane-*block*-styrene] (**PTMVES-*b*-PS**). Made from PTMVES (GPC M_n = 8200 g/mol, M_w/M_n = 1.16), GPC (diblock) M_n = 75200 g/mol, M_w/M_n = 1.53.

Poly(trimethyl((4-vinylphenyl)ethynyl)silane)-*block*-poly(isothiocyanato styrene) (**PTMSAS-*b*-PITS**). Made from PTMVES (GPC M_n = 14000 g/mol, M_w/M_n = 1.29), GPC (diblock) M_n = 23500 g/mol, M_w/M_n = 1.64. ^{1}H-NMR, 300 MHz, $CDCl_3$ δ (ppm) = 7.12, 6.91, 6.35 (m, 4 H, Ar-H); 1.83-1.23 (m, 3 H, backbone); 0.24 (bs, 4.3 H, Si-CH_3).

General Procedure for RAFT Polymerization

An example is given: 500 mg (3.1 mmol) of isothiocyanato styrene, 30 mg (0.123 mmol) of benzyl dithiobenzoate and 2.5 mg (15.2 μmol) of AIBN were heated for 3.5 hours to 60°C. THF was added and the polymer was precipitated into methanol. GPC M_n = 8500 g/mol, M_w/M_n = 1.37.

Poly(isothiocyanato styrene-*block*-ethyl acrylate) (**PITS-*b*-EA**). 53 mg of PITS prepared by RAFT (M_n = 8500 g/mol, M_w/M_n = 1.37) were heated with

100 mg (1 mmol) of ethyl acrylate, 1 mg (6.1 μmol) of AIBN and 2 mL of THF to 60°C for 4 days. After prepipitating the mixture twice into methanol, 60 mg of polymer were obtained. GPC M_n = 17000 g/mol, M_w/M_n = 1.93; ^{1}H-NMR, 300 MHz, $CDCl_3$ δ (ppm) = 7.6 - 6.2 (m, Ar-H), 4.2-4.0 (m, -O$C\underline{H}_2CH_3$), 2.4-0.8 (m,backbone).

Poly(*p*-azidomethyl styrene)

1.87 g (12.4 mmol) of *p*-chloromethyl styrene and 4 mg (24.4 μmol) of AIBN were stirred at 90°C for 74 min. The resulting polymer was precipitated into methanol and dried. 970 mg (52%) of poly(*p*-chloromethyl styrene) were obtained. ^{1}H-NMR, 300 MHz, $CDCl_3$ δ / ppm = 6.70-7.20 (m, 2 H, Ar-H), 6.20-6.60 (m, 2 H, Ar-H), 4.20-4.60 (m, 2 H, Ar-CH_2Cl), 1.20-1.90 (m, 3 H, -$CHCH_2$-).

400 mg (2.7 mmol monomer units) of this polymer were stirred at room temperature with 0.34 g (5.2 mmol) of sodium azide in 24 mL of acetonitrile / DMSO (1:1) for 4.5 hours. The polymer was precipitated into methanol and dried. GPC M_n = 186000 g/mol, M_w/M_n = 1.79; ^{1}H-NMR, 300 MHz, $CDCl_3$ δ / ppm = 6.70-7.20 (m, 2 H, Ar-H), 6.20-6.60 (m, 2 H, Ar-H), 3.90-4.00 (m, 2 H, Ar-CH_2N_3), 1.20-1.90 (m, 3 H, -$CHCH_2$-).

Layer Deposition

Prior to polymer immobilization, glass slides were cleaned with water / conc. sulfuric acid / 35% H_2O_2 (5:1:1) at 70°C for 20 minutes and then rinsed with ultrapure water. One slide was kept as a reference for UV-vis measurements. Glass slides to be dip coated with polymers were consequently dipped into the following solutions. (1) polyethylene imine (Aldrich, M_n – 25 kg/mol) in ethanol, 1 mg/mL, 20 min at room temperature. (2) Diblock copolymer PTMVES-*b*-PITS in THF, 1 mg/mL, for 20 min at room temperature. (3) Tetrabutylammonium fluoride in THF, 0.3 mg/ mL for 20 min at room temperature. (4) Poly(*p*-azidomethyl styrene) in THF at 5 mg/mL with a trace of Cu(I)Br. UV-vis measurements were recorded after 20 min, after 45 min, and after 12 hours at room temperature. Glass slides were thoroughly washed 3 times with THF after each immersion.

Conclusions

We could demonstrate the synthesis of a new monomer 4-isothiocyanato styrene (ITS), which in combination with trimethyl((4-vinylphenyl)ethynyl)silane (TMVES) could be polymerized under controlled radical polymerization conditions leading essentially to orthogonally reactive diblock copolymers. These block copolymers could then successfully be employed in the preparation of covalently bound LbL thin films that allowed for a sequential control of layer deposited, thereby providing the first step to a precise control of internal thin film control at the nanometer scale.

Acknowledgments

R. Rieger is gratefully acknowledged for support in the experimental work. This research was partly supported by WCU (World Class University) program through the National Research Foundation of Korea funded by the Ministry of Education, Science and Technology (R31-10013).

References

1. Kim, J. K.; Lee, J. I.; Lee, D. H. *Macromol. Res.* **2008**, *16*, 267–292.
2. Sumerlin, B. S.; Vogt, A. P. *Macromolecules* **2010**, *43*, 1–13.
3. Theato, P. *J. Polym. Sci., Part A: Polym. Chem.* **2008**, *46*, 6677–6687.
4. Fournier, D.; Hoogenboom, R.; Schubert, U. S. *Chem. Soc. Rev.* **2007**, *36*, 1369–1380.
5. Lutz, J. F. *Angew. Chem., Int. Ed.* **2007**, *46*, 1018–1025.
6. Malkoch, M.; Thibault, R. J.; Drockenmuller, E.; Messerschmidt, M.; Voit, B.; Russell, T. P.; Hawker, C. J. *J. Am. Chem. Soc.* **2005**, *127*, 14942–14949.
7. Kessler, D.; Nilles, K.; Theato, P. *J. Mater. Chem.* **2009**, *19*, 8184–8189.
8. Ghosh, S.; Basu, S.; Thayumanavan, S. *Macromolecules* **2006**, *39*, 5595–5597.
9. Yang, S. K.; Weck, M. *Macromolecules* **2008**, *41*, 346–351.
10. Uenishi, K.; Sudo, A.; Endo, T. *J. Polym. Sci., Part A: Polym. Chem* **2009**, *47*, 6750–6757.
11. Messerschmidt, M.; Millaruelo, M.; Komber, H.; Häussler, L.; Voit, B. *Macromolecules* **2008**, *41*, 2821–2831.
12. Burd, C.; Weck, M. *J. Polym. Sci., Part A: Polym. Chem.* **2008**, *46*, 1936–1944.
13. Nilles, K.; Theato, P. *J. Polym. Sci., Part A: Polym. Chem.* **2010**, *48*, 3683–3692.
14. Decher, G.; Hong, J. D.; Schmitt, J. *Thin Solid Films* **1992**, *210–211*, 831–835.
15. Decher, G. *Science* **1997**, *277*, 1232–1237.
16. Lvov, Y. M.; Lu, Z.; Schenkman, J. B.; Zu, X.; Rusling, J. F. *J. Am. Chem. Soc.* **1998**, *120*, 4073–4080.
17. Wood, K. C.; Chuang, H. F.; Batten, R. D.; Lynn, D. M.; Hammond, P. T. *Proc. Natl. Acad. Sci. U.S.A.* **2006**, *103*, 10207–10212.
18. Kim, B.-S.; Park, S. W.; Hammond, P. T. *ACS Nano* **2008**, *2*, 386–392.
19. Berg, M. C.; Zhai, L.; Cohen, R. E.; Rubner, M. F. *Biomacromolecules* **2006**, *7*, 357–364.
20. Farhat, T. R.; Hammond, P. T. *Adv. Funct. Mater.* **2005**, *15*, 945–954.
21. Jiang, S. P.; Liu, Z.; Tian, Z. Q. *Adv. Mater.* **2006**, *18*, 1068–1072.
22. Nandivada, H.; Jiang, X.; Lahann, J. *Adv. Mater.* **2007**, *19*, 2197–2208.
23. Stockton, W. B.; Rubner, M. F. *Macromolecules* **1997**, *30*, 2717–2725.
24. Kharlampieva, E.; Sukhishvili, S. A. *Polym. Rev.* **2006**, *46*, 377–395.
25. Kohli, P.; Blanchard, G. J. *Langmuir* **2000**, *16*, 4655–4661.

26. Such, G. K.; Quinn, J. F.; Quinn, A.; Tjipto, E.; Caruso, F. *J. Am. Chem. Soc.* **2006**, *128*, 9318–9319.
27. Bergbreiter, D. E.; Liao, K.-S. *Soft Matter* **2009**, *5*, 23–28.
28. Buck, M. E.; Zhang, J.; Lynn, D. M. *Adv. Mater.* **2007**, *19*, 3951–3955.
29. Seo, J.; Schattling, P.; Lang, T.; Jochum, F.; Nilles, K.; Theato, P.; Char, K. *Langmuir* **2010**, *26*, 1830–1836.
30. Tsuda, K.; Ishizone, T.; Hirao, A.; Nakahama, S. *Macromolecules* **1993**, *26*, 6985–6991.
31. Shimo, N.; Nakashima, N.; Yoshihara, K. *Chem. Phys. Lett.* **1986**, *125*, 303–306.
32. Li, T.; Zhou, C.; Jiang, M. *Polym. Bull.* **1991**, *25*, 211–216.

Chapter 4

Exploiting Architecture and Composition of Side-Chain Liquid Crystalline Polymers for Shape Memory Applications

S.-k. Ahn,[1] P. Deshmukh,[2] and R. M. Kasi[1,2,*]

[1]Polymer Program, Institute of Materials Science, [2]Department of Chemistry, University of Connecticut, Storrs, CT 06269
[*]Corresponding author: kasi@ims.uconn.edu

We have investigated methods to improve shape memory performance by exploiting architecture and composition of side-chain liquid crystalline polymers (SCLCPs). Three different bicyclo[2.2.1]hept-2-ene monomers bearing cholesteryl ester (NBCh), poly(ethylene glycol) (NBPEG) and butyl acrylate (NBBA) are synthesized. All the polymers are synthesized by ring-opening metathesis polymerization (ROMP) of the above monomers. While the homopolymer comprises only NBCh, co- and terpolymers contain NBCh as the major component and NBBA and/or NBPEG as minor components, respectively. NBCh is used so that its bioactivity and liquid crystalline properties can be harnessed for biomedical applications. NBBA and/or NBPEG are introduced to tailor physical and thermal properties of the polymers. Thermo-mechanical properties as well as shape memory properties in these polymers are compared, and strategies for outstanding shape memory performance are discussed.

Introduction

Shape memory polymers (SMPs) are a class of stimuli-sensitive polymers which can recover a desired permanent shape from a temporarily programmed shape when exposed to stimuli such as heat, light, pH, moisture, magnetic and electrical field. The unique feature of SMPs that recovers a pre-defined shape even after large deformation up to several hundred percent allows the SMPs to be adapted for applications in smart functional devices including heat-shrinkable tubes (*1*), self-healing plastics (*2*), actuators (*3*) and textiles (*4*). In addition, SMPs can also serve as active substrates to fabricate recyclable elastomeric stamp (*5*), to create artificial surface wrinkles (*6*) and to guide cell growth (*7*). Furthermore, the use of biodegradable as well as biocompatible polymers enables the application of SMPs in multi-functional biomedical devices such as apparatuses for minimally invasive surgery (*8*), self-expandable stents (*9*), tissue engineering scaffolds (*10*), and controlled drug release vehicles (*11*).

Major class of current SMPs utilizes heat as a stimulus to trigger shape memory effect (SME). These thermally-induced SMPs are based on exploiting one or more phase transition temperature (T_{trans}) such as glass transition (T_g), melting temperature (T_m) or liquid crystalline clearing temperature (T_{cl}). Generally, the SME can be considered as combination of two processes called programming and recovery process. The programming process is a step to create a temporary shape by applying external force and temperature to a predefined permanent shape. During this process, the temporary shape can be typically obtained by deformation of the sample at the elevated temperature above T_{trans} followed by cooling the sample to below T_{trans} under constant stress to freeze the deformed shape. In the recovery process, the permanent shape is recovered by reheating the SMP above T_{trans} under stress-free condition. The entropy elasticity is the driving force to recover the permanent shape from the temporary shape.

There are two key requirements for SME: permanent and temporary cross-links. Permanent cross-links determine the original shape of SMP during a fabrication procedure. Also, the disentanglement of polymeric chain can be prevented by the permanent cross-links, thereby irreversible creep responses during shape memory cycles (SMCs) are suppressed leading to improved shape memory performance. Permanent cross-links can be introduced through chemical or physical cross-links. Chemically, the polymeric chain can be interconnected by thermal or photo curing in the presence or absence of initiators. Physically, the permanent cross-links are commonly obtained by phase-separated block copolymer strategy where switching-segment domain (with low thermal transition) is responsible for activating SME, while hard-segment domain (with high thermal transition) stabilizes the permanent shape (*12*). Non-covalent interaction such as ionic clusters can also form physical cross-links to preserve the permanent shape (*13*, *14*). Temporary cross-links are physically formed reversible networks which typically rely upon vitrification below T_g or crystallilzation below T_m. Vitrified polymeric segments usually exhibits better shape fixing capability than crystallized polymeric phase since the latter can be influenced by the crystallization kinetics. Liquid crystalline ordering, particularly for smectic mesophase, is also shown to temporarily hold the shape (*5*, *15*).

Recently, we reported multiple SMEs by using T_g and T_{cl} of a side-chain liquid crystalline (SCLC) type terpolymer network that shows smectic A (SmA) mesophase comprising cholesteryl mesogens (*15*). Permanent shape is stabilized by chemical cross-links, and temporary shape is attained by either vitrification or smectic layer ordering depending on type of SMC. More recently, we have also demonstrated that SmA polymorphism in nanometer scale which originates from different extent of interdigitated structure of cholesteryl mesogens determines

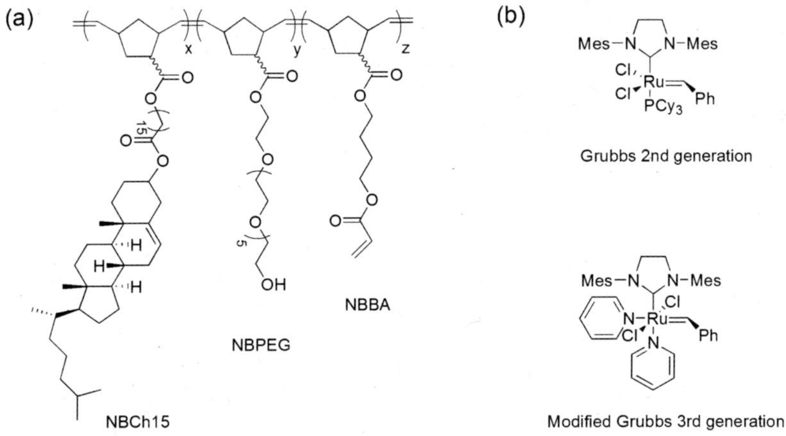

Figure 1. (a) Chemical structure of polymers: homo-, co- and terpolymer are obtained by varying the molar composition (x, y and z) of each monomer: i) $x = 100$ and $y = z = 0$ for homopolymer, ii) $x \neq z \neq 0$ and $y = 0$ for copolymer, iii) $x \neq y \neq z \neq 0$ for terpolymer. (b) Ruthenium based catalysts used in polymerization (Mes = 2,4,6-trimethylphenyl).

macroscopic strain responses (*16*). This interdigitation based thermostrictive property can be used to develop actuators.

In this study, we have systematically investigated methods to attain excellent shape memory performance by the choice and composition of different comonomers using the T_g of polymer backbone as a trigger. A library of homo-, co- and terpolymer have been synthesized by varying the composition of three bicyclo[2.2.1]hept-2-ene based monomers which comprise 1) cholesteryl ester, 2) amorphous poly(ethylene glycol) (PEG) and 3) butyl acrylate (BA) side-chains through ring-opening metathesis polymerization (ROMP). For co- and terpolymers with higher composition of cholesteryl ester, the resulting polymer is expected to show a block-like distribution due to strong interaction between hydrophobic cholesteryl group. The thermo-mechanical properties and T_g based shape memory properties in these polymers are discussed. Unlike other conventional T_g based SMPs, the additional bioactivity and alignment capability of liquid crystalline cholesterol mesogens (*17–19*) imparts multi-functionality to the polymers. These features together with the shape memory properties, makes these materials suitable for biomedical and tissue engineering applications.

Results and Discussion

Synthesis of Polymers

Figure 1 shows chemical structures of SCLC type polymers and ruthenium based catalysts (Grubbs catalyst second (G2nd) and modified Grubbs catalyst third generation (mG3rd)) used for ROMP. Three different norbornene derivative monomers, 5-{*15*-(cholesteryloxycarbonyl)-pentadecyloxycarbonyl}-bicyclo[2.2.1]hept-2-ene (NBCh15), poly(ethylene glycol) functionalized norbornene (NBPEG) and 5-(acryloyl butoxycarbonyl)-bicyclo[2.2.1]hept-2-ene (NBBA) are prepared (*15, 20*). The each monomer has a unique function in the polymers, where 1) NBCh15 provides LC ordering, 2) amorphous NBPEG acts as an internal plasticizer to control thermal transitions and mechanical property, and 3) NBBA affords reactive sites that can be cross-linked by either heat or light after polymerization. Depending on the choice of each monomer, three different types of polymers including homo- (HP15), co- (CP15) and terpolymers (TP15) are prepared. All these polymers are designed to contain cholesteryl moieties as the major components by which further biofuntionality as well as LC ordering can be provided.

Table I summarizes composition and molecular properties of homo-, co- and terpolymers determined using proton nuclear magnetic resonance (^{1}H NMR) and gel permeation chromatography (GPC). The composition of each co- and terpolymer is varied by changing monomer feed ratio and this allows for investigating the influence of composition and architecture on thermal and final shape memory properties. We have used two different types of Grubbs catalysts (G2nd and mG3rd). The mG3rd is prepared by adding trace amount of pyridine to the commercially available G2nd, followed by precipitating in excess pentane (*21*). When the mG3rd is used, polymerization is completed within shorter time (10 min) compared to G2nd (1.5 h) and narrower polydispersity indices (PDI) are obtained.

Thermal and Mechanical Properties

Thermal transitions of linear polymers determined by second heating curves of differential scanning calorimetry (DSC) are shown in Figure 2. A homopolymer (HP15(100)) exhibits a step-like transition of T_g and a sharp T_{cl} attributed to the large decoupling of motion between cholesteryl side-chains and polymer main-chains. When comonomers (NBBA and NBPEG) are introduced to the system, both T_g and T_{cl} clearly shift to lower temperatures indicating both PEG and BA act as internal plasticizers. A block-like distribution is expected for co- and terpolymer with higher molar composition of cholesteryl groups (CP15(70) and TP15(80)) since distinct T_g and sharp T_{cl} are observed. However, increasing amounts of NBBA in the copolymers (CP15(50) and CP15(30)) broadens both T_g and T_{cl} by diluting interactions of cholesteryl groups.

The chemical networks which determine a permanent shape of SMPs are formed by thermally curing the polymers inside vacuum oven at 120 °C for two days. Gel fraction values determined by Soxhlet extraction using THF shows reasonable extent of cross-linking. After curing, T_g slightly increases, while the enthalpy of T_{cl} decreases. Inclusion of more NBBA in the copolymers leads to higher T_g and lower values of enthalpy in T_{cl} since covalent networks restrict both the segmental motion of polymeric chain and the LC ordering from cholesteryl moiety. The gel fraction (G) values and thermal transitions of linear and cross-linked polymers are listed in Table II.

Table I. Synthesis of SCLC type polymers

polymer[a]	catalyst[b]	mol %[c]			M_n (kDa)[d]	PDI[d]
		NBCh15	NBPEG	NBBA		
HP15(100)	G2nd	100	-	-	98	1.50
CP15(70)	mG3rd	70.4	-	29.6	38	1.23
CP15(50)	mG3rd	49.5	-	50.5	28	1.27
CP15(30)	mG3rd	29.6	-	70.4	24	1.34
TP15(80)	G2nd	80.5	15.5	4.0	97	1.54

[a] Values in parenthesis denote mol % of cholesteryl monomer (NBCh15). [b] Types of [Ru] derivative catalyst. [c] Calculated from ^{1}H NMR. [d] Obtained from GPC with THF as eluent and calibrated with PS standards.

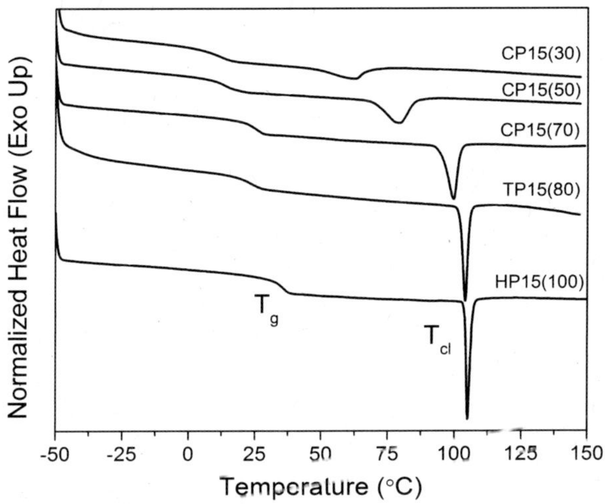

Figure 2. Second heating curves of DSC measurement with scanning rate of 10 °C/min.

The stress-strain measurements on polymers have been conducted at 60 °C (Figure 3), which will be used as the deformation temperature for T_g based SMC. These experiments allow us to estimate how large each polymer can be extended during the deformation step in SMC. The elongation at break (ε_b) for each polymer at 60 °C is listed in Table II. Homopolymer (HP15(100)) is shown to be highly stretchable (ε_b = 322%) arising from the ability of adopting different conformation under stress by long and flexible methylene spacer. The XL-CP15 (70) and XL-CP15 (50) shows less elongation than homopolymer. Moreover, increased amounts of NBBA in the copolymers reduces the strain significantly to about 76 %, suggesting that increased number of chemical networks prevents the polymeric segment from deformation. The XL-TP15(80) exhibits outstanding ε_b of ~300% mainly because of the softening effect due to incorporation of amorphous PEG, which also enhances the toughness of XL-TP15(80). The 5 mol % of NBBA allows the terpolymer to be thermally cured, resulting in a lightly cross-linked material. Thus, for a highly deformable and toughened shape memory material, it would be desirable to tune thermal and physical properties by varying the ratio between NBCh15 and NBPEG, while retaining low amounts of cross-linkable NBBA moieties.

The linear viscoelastic properties of the materials are investigated by dynamic mechanical analysis (DMA) in tensile mode. Figure 4 compares the results of the homopolymer, cross-linked co- and terpolymers. The typical storage modulus (~1 GPa) for glassy polymers is found below T_g, which dramatically drops to a value less than 2 MPa upon heating above T_g. Further increase of temperature to T_{cl}, leads a flow of HP15(100), while chemically cross-linked co- and terpolymers exhibit rubbery plateaus. The higher loading of NBBA increases the modulus of rubbery plateaus from 0.05 MPa for XL-TP15(80) to 0.2 MPa for XL-CP15(70) and 1.4 MPa for XL-CP15(50), which indicates larger amount of cross-linking density. The tan δ, associated with the relaxation of polymeric segments, shows two or three characteristic peaks in these polymers. The relaxation peaks for T_g and broad weak peaks corresponding to T_{cl} are found in all cross-linked polymers. Interestingly, we have observed additional peaks in tan δ trace between T_g and T_{cl} although it is hard to detect in these transitions in XL-CP15(50). We speculate that these peaks might be due 1) dynamic soft elasticity (*22*) and 2) relaxation from smectic layer (*23*, *24*). The first premise originates from energy dissipation during polydomain to monodomain transition upon loading which results in the reorientation of LC domain. The second hypothesis is based on the large decoupling of motion between polynorbornene main-chain and cholesteryl side-chains. In other words, 15 methylene spacer in NBCh15 distinctly separates movement of cholesteryl side-chains from main-chain so that the relaxation of smectic layer can be independently observed from tan δ trace beyond relaxation of T_g of polynorbornene main-chain. However, further rheological studies are required to thoroughly understand the relaxation behaviors in these polymers.

Shape Memory Properties

One-way (1W) shape memory properties of linear homopolymer and cross-linked co- and terpolymers are explored by means of a cyclic thermo-mechanical analysis using DMA. Temperature ranges for the SMC is selected as 0 - 60 °C which includes $T_g \pm 20$ °C, but excludes T_{cl}. This T_g based 1W-SMCs is programmed by 1) heating samples to 60 °C ($T > T_g$) and deforming the samples by ramping a load, 2) cooling the samples to 0 °C ($T < T_g$) under constant load, 3) unloading at 0 °C and 4) reheating the samples to 60 °C. The first three steps are required obtain a temporary shape and the last reheating step is a recovery procedure. We have recently published T_{cl} based 1W-SMCs wherein the strain profile is strongly influenced by the extent of interdigitation within smectic layers formed by cholesteryl mesogens (*16*).

The quantitative shape memory properties (strain, shape fixity (R_f) and shape recovery ratio (R_r)) are obtained through the analysis of the T_g based 1W-SMCs. R_f represents for the ability of a material to store mechanical deformation after

Table II. Thermal and mechanical properties of linear and cross-linked polymers

polymer	*G* (%)[a]	T_g (°C)[b]	T_{cl} (°C)[b]	ΔH for T_{cl} (J/g)[b]	ε_b (60°C) (%)[c]
HP15(100)	-	35.6	104.6	5.14	322
CP15(70)	-	25.3	99.6	5.9	-
XL-CP15(70)	74.1	27.9	99.6	5.0	126
CP15(50)	-	14.0	79.5	5.8	-
XL-CP15(50)	84.5	23.3	81.5	2.5	30
CP15(30)	-	8.4	62.6	3.7	-
XL-CP15(30)	92.1	15.3	60.2	0.8	n/a[d]
TP15(80)	-	23.0	104.1	5.5	-
XL-TP15(80)	89.0	22.5	103.5	5.4	298

[a] G: gel fraction determined from mass ratio between before and after Soxhlet extraction by THF as solvent. [b] Thermal properties are determined from second heating cycles of DSC thermogram. [c] ε_b represents elongation at break of polymers at 60 °C. [d] The sample is not investigated due to processability issue.

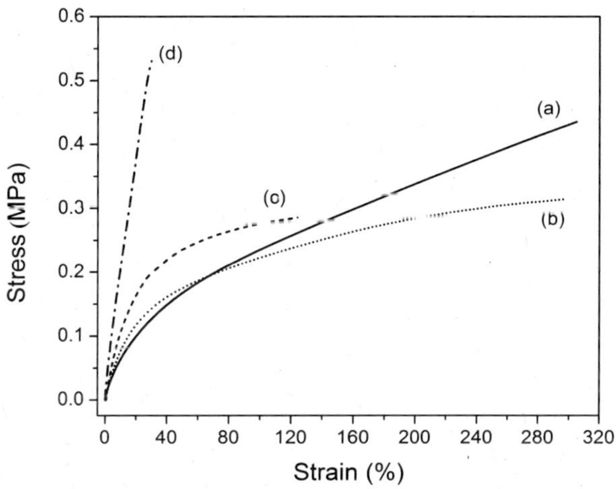

Figure 3. Stress-strain curves at 60°C under 1N/min of force ramp: (a) HP15(100), (solid), (b) XL-TP15(80), (short dot), (c) XL-CP15(70), (short dash) and (d) XL-CP15(50), (dash dot).

unloading and R_r describes the ability of a material to recover mechanical deformation by regaining entropy elasticity. R_f and R_r are calculated from the following equations (*25*),

$$R_f(\%) = \frac{\varepsilon_u(N) - \varepsilon_p(N-1)}{\varepsilon_l(N) - \varepsilon_p(N-1)} \times 100 \qquad R_r(\%) = \frac{\varepsilon_u(N) - \varepsilon_p(N)}{\varepsilon_u(N) - \varepsilon_p(N-1)} \times 100$$

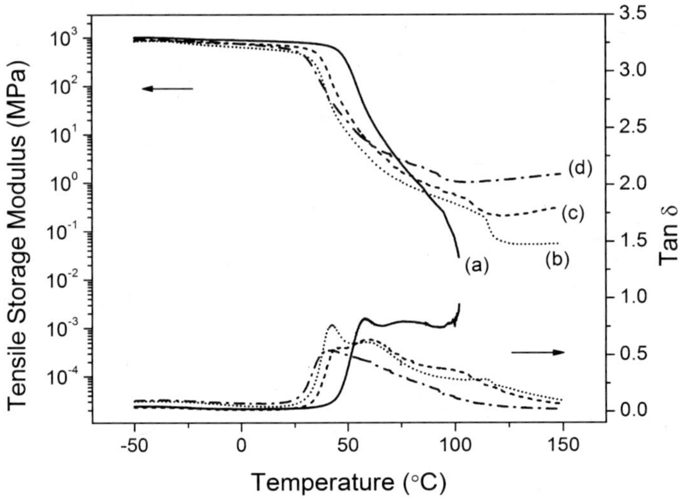

Figure 4. Linear viscoelastic properties: (a) HP15(100), (solid), (b) XL-TP15(80), (short dot), (c) XL-CP15(70), (short dash) and (d) XL-CP15(50), (dash dot).

where, ε_u is the strain after unloading, ε_l is the strain with load before unloading, ε_p is the strain after recovery and N is the number of cycles.

Figure 5 describes the three continuous 1W-SMCs of three different polymers. In Figure 5a, the strain increment of HP15(100) in the deformation step increases with cycle number. This implies that the polymeric chain slides against each other more with increasing SMCs due to the lack of permanent physical or chemical networks. The entanglement of polymeric chain which can be another type of physical cross-link is not strong enough to prevent slippage. The XL-CP15(70) in Figure 5b, does not show an irreversible creep behavior unlike HP15(100) because of the large number of chemical junctions formed by acrylate group. However, the strain increment in the deformation step decreases with increasing cycle number, which suggests that the material becomes stiffer during the SMCs. This is possibly because not all of the NBBA are cross-linked before SMC and additional curing of acrylate group in NBBA occurs during SMC. To confirm our speculation, we have analyzed DSC of samples that have been through SMCs. T_g barely changes before and after SMCs, however, enthalpy of T_{cl} decreases about 36 % indicating further curing during SMCs that stiffens the material and dilutes liquid crystalline ordering. More importantly, XL-TP15(80) (Figure 5c) exhibits highly reproducible SMCs with large strain (~140 %). In other words, the introduction of 5 mol % of NBBA to terpolymer is sufficient to prevent polymeric chain from sliding and amorphous PEG side-chain provides toughness of the material to be able to withstand elongation.

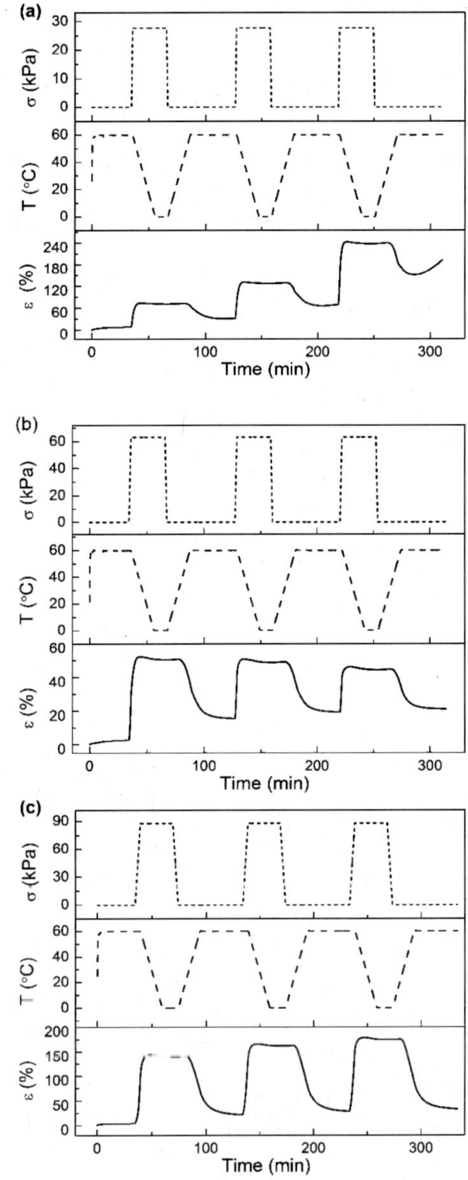

Figure 5. T_g based one-way shape memory cycles: (a) HP15(100), (b) XL-CP15(70) and (c) XL-TP15(80).

Table III. Quantitative shape memory properties during T_g based 1W-SMCs

polymer	*cycle no.*	*ε (%)*	*R_f (%)*	*R_r (%)*
	1	62.3	100.0	65.0
HP15(100)[a]	2	96.0	100.0	64.1
	3	167.9	100.0	51.0
	1	47.7	99.9	73.0
XL-CP15(70)	2	33.2	99.8	88.8
	3	25.1	99.8	92.8
	1	10.1	99.2	83.6
XL-CP15(50)	2	6.9	98.9	93.1
	3	5.5	98.7	95.5
	1	136.3	99.9	87.1
XL-TP15(80)	2	140.4	99.9	95.2
	3	146.1	99.9	96.8

[a] R_r is calculated by taking the minimal strain during 40 minutes of annealing

Table III summarizes the strain, R_f and R_r of homo-, co- and terpolymers during SMCs. In all cases, shape fixing is quantitative indicating that vitrified polymeric chains below T_g efficiently freezes the temporarily elongated strain. As depicted in Figure 5a, the strain of HP15(100) continuously increases in subsequent cycles up to 168 % due to the absence of permanent network. Shape recovery is also weak which could be attributed to the plastic deformation resulting from unconstrained, sliding polymeric chains. XL-CP15(70) shows reduced strain in subsequent cycles due to the extra thermal curing while shape recovery improves with increasing cycle number. A similar trend in strain and R_r is also observed from XL-CP15(50). However, increased number of cross-links in the material resulting from high loading of NBBA permits only limited elongation. When 15 mol % of amorphous NBPEG is incorporated to the system in addition to the 5 mol % of NBBA, the resulting XL-TP15(80) exhibits excellent shape memory properties including large elongation and nearly complete shape fixing and recovery. In XL-TP15(80), NBPEG improves the toughness of the material, while cross-linkable NBBA moiety efficiently restores the shape by preventing sliding of polymeric chain. Thus, XL-TP15(80) is the optimized architecture and composition to attain highly stretchable SMP with excellent shape fixing and shape recovery properties among the series of homo-, co- and terpolymers bearing SCLC cholesterol molecules. As a final thought, T_g or T_{cl} can be tailored by varying the composition of the terpolymer using different amounts of NBCh15 and NBPEG, while still retaining about 5 mol % of NBBA in the monomer feed (*15*).

Summary

We have synthesized a series of SCLC type linear and cross-linked homo-, co- and terpolymers, and investigated their thermomechanical and shape memory properties. The thermal and mechanical properties of the polymers are tuned by varying the monomer feed ratio and cross-linking density. In co- and terpolymers containing large amount of cholesteryl monomer, a block-like distribution is expected due to strong interaction between cholesteryl groups. Excellent shape memory performance (large strain, repeatable SMCs, high shape fixing and shape recovery ratio) are achieved from a lightly cross-linked terpolymer. This can be attributed to the presence of comonomers comprising amorphous PEG and cross-linkable BA side-chains in the terpolymer. The bioactivity and LC alignment of cholesteryl groups along with tunable mechanical, thermal and shape memory properties allows this new class of polymer to serve as a promising candidate for applications in biotechnology.

Experimental

Materials

Synthesis and characterization of three monomers (NBCh15, NBPEG and NBBA) and the general protocol to synthesize all polymers are reported in our previous publications (*15*, *20*).

Molecular Characterization

^{1}H NMR spectra (Bruker DMX 500MHz NMR spectrometer) are recorded with tetramethylsilane (TMS) as an internal standard and deuterated chloroform ($CDCl_3$) as lock solvent. The molar composition of co- or terpolymers is determined by comparing the integration ratio of characteristic NMR peaks which correspond to each monomer. Molecular weight and PDI are determined by using a Waters 1515 GPC coupled with a PL-ELS100 (evaporative light scattering) and a Waters 2487 dual wavelength absorbance detector with THF as eluent. Polystyrene is used to construct calibration curves.

Thermal and Mechanical Analyses

Thermal properties of polymers are obtained from DSC (Q20, TA instrument) under a nitrogen purge. Samples are preheated to 150 °C to remove thermal history and second heating cycles are reported for T_g, T_{cl} and ΔH using Universal Analysis software. Thermomechanical properties are studied by DMA 2980 (TA instrument) with cryo accessories attached. For stress-strain measurement, the samples are thermally equilibrated at 60 °C for 5 min, followed by ramping a force at a rate of 1 N/min until samples break. The temperature dependent experiment is conducted using a tension film mode. Preload force of 0.01 N, a heating rate of 3 °C/min, force track (ratio of static to dynamic force) of 125 % and an oscillation frequency of 1 Hz are selected.

Thermal Curing

The powder type samples are compression molded at 20-30 °C above T_g of each polymer to obtain uniform films with defined thickness. Afterward, the films placed between two hot plates are thermally cured inside vacuum oven at 120 °C for 48h. The gel fraction analyses on cured samples are carried out by Soxhlet extraction using THF for 24 h. The gel fraction values are determined by the ratio between initial mass before extraction (m_i) and dried mass after extraction (m_d),

$$G(\%) = \frac{m_d}{m_i} \times 100$$

Shape Memory Characterization

Cyclic thermomechanical analyses are performed on a DMA 2980 operated in controlled force mode and temperature range of 0-60 °C. One complete SMC consists of four consecutive steps. Prior to start SMC, the sample is relaxed at 60 °C ($T > T_g$) for 30 min to be thermally equilibrated. In step 1, the sample is uniaxially elongated by ramping a force from preload 0.01 N to certain value at a rate of 0.05 N/min (deformation). In step 2, the sample is cooled at a rate of 3 °C/min to 0 °C ($T < T_g$) to solidify the deformed sample under constant load (cooling). In step 3, the force applied to the sample is reduced to preload value (0.01 N) at a rate of 0.05 N/min (unloading and shape fixing). In the final step, sample is reheated at a rate of 3 °C/min to 60 °C and equilibrated for additional 40 min to recover any residual strain (recovery). The SMC consisting of four steps is repeated at least two more times on the same sample.

Acknowledgments

Financial support was provided by the University of Connecticut new-faculty start up fund, University of Connecticut Research Foundation faculty grant, and NSF CAREER Award to R.M.K. (DMR-0748398). Central instrumentation facilities in the Institute of Materials Science and Chemistry Department at the University of Connecticut are acknowledged.

References

1. Charlesby, A. *Atomic Radiation and Polymers*; Pergamon Press: Oxford, 1960.
2. Garcia, M. E.; Lin, Y.; Sodano, H. A. *J. Appl. Phys.* **2010**, *108*, 093512.
3. Gall, K.; Kreiner, P.; Turner, D.; Hulse, M. *J. Microelectromech. Syst.* **2004**, *13*, 472–483.
4. Hu, J.; Chen, S. *J. Mater. Chem.* **2010**, *20*, 3346–3355.
5. Burke, K. A.; Mather, P. T. *J. Mater. Chem.* **2010**, *20*, 3449–3457.
6. Xie, T.; Xiao, X.; Li, J.; Wang, R. *Adv. Mater.* **2010**, *22*, 4390–4394.
7. Davis, K. A.; Burke, K. A.; Mather, P. T.; Henderson, J. H. *Biomaterials* **2011**, *32*, 2285–2293.

8. Lendlein, A.; Langer, R. *Science* **2002**, *296*, 1673–1676.
9. Xue, L. A.; Dai, S. Y.; Li, Z. *Biomaterials* **2010**, *31*, 8132–8140.
10. Thornton, A. J.; Alsberg, E.; Albertelli, M.; Mooney, D. J. *Transplantation* **2004**, *77*, 1798–1803.
11. Wischke, C.; Neffe, A. T.; Steuer, S.; Lendlein, A. *J. Controlled Release* **2009**, *138*, 243–250.
12. Lendlein, A.; Kelch, S. *Angew. Chem., Int. Ed.* **2002**, *41*, 2034–2057.
13. Weiss, R. A.; Izzo, E.; Mandelbaum, S. *Macromolecules* **2008**, *41*, 2978–2980.
14. Xie, T. *Nature* **2010**, *464*, 267–270.
15. Ahn, S.-k.; Deshmukh, P.; Kasi, R. M. *Macromolecules* **2010**, *43*, 7330–7340.
16. Ahn, S.-k.; Deshmukh, P.; Gopinadhan, M.; Osuji, C. O.; Kasi, R. M. *ACS Nano* 2011, in press.
17. Hwang, J. J.; Iyer, S. N.; Li, L.-S.; Claussen, R.; Harrington, D. A.; Stupp, S. I. *Proc. Natl. Acad. Sci. U.S.A.* **2002**, *99*, 9662–9667.
18. Klok, H.-A.; Hwang, J. J.; Iyer, S. N.; Stupp, S. I. *Macromolecules* **2002**, *35*, 746–759.
19. Zhou, Y.; Briand, V.; Sharma, N.; Ahn, S.-k.; Kasi, R. M. *Materials* **2009**, *2*, 636–660.
20. Ahn, S.-k.; Nguyen Le, L. T.; Kasi, R. M. *J. Polym. Sci., Part A: Polym. Chem.* **2009**, *47*, 2690–2701.
21. Love, J. A.; Morgan, J. P.; Trnka, T. M.; Grubbs, R. H. *Angew. Chem., Int. Ed.* **2002**, *41*, 4035–4037.
22. Qin, H.; Mather, P. T. *Macromolecules* **2009**, *42*, 273–280.
23. Mano, J. F.; Gómez Ribelles, J. L. *Macromolecules* **2003**, *36*, 2816–2824.
24. Craig, A. A.; Winchester, I.; Madden, P. C.; Larcey, P.; Hamley, I. W.; Imrie, C. T. *Polymer* **1998**, *39*, 1197–1205.
25. Behl, M.; Bellin, I.; Kelch, S.; Wagermaier, W.; Lendlein, A. *Adv. Funct. Mater.* **2009**, *19*, 102–108.

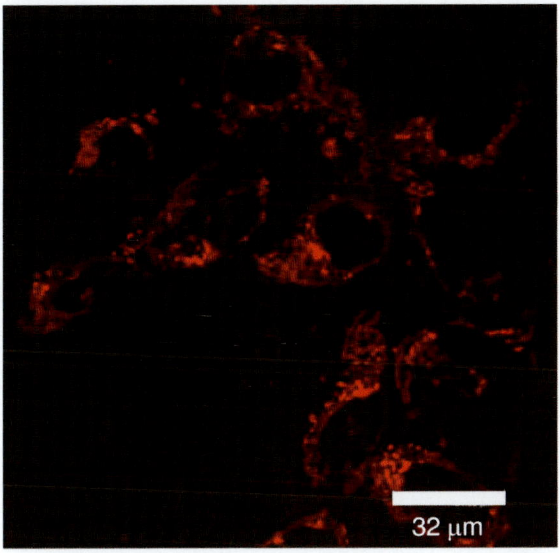

Figure 8. Confocal microscopy image of HEp-2 carcinoma cells after a 22h incubation period.

Chapter 5

Nonconventional Elements in Block Copolymers

E. Bryan Coughlin* and Yoan Simon

Department of Polymer Science and Engineering, University of Massachusetts, 120 Governors Drive, Amherst, MA 01003
***coughlin@mail.pse.umass.edu**

The conventional compositions of block copolymers most often employ the elements C, H, and perhaps O or N. We are expanding the elemental composition of block copolymer by using Boron containing monomers. The synthesis of a novel silyl-protected oxanorbornene imide carborane (SONIC) monomer as well as its homo and co-polymerization by ring-opening metathesis polymerization (ROMP) to obtain amphiphilic diblock copolymers is described. The subsequent functionalization with oligo ethylene oxide side chains and fluorescent dye labeling of these block polymers, as well as their solution behavior have been investigated. Initial biological studies have shown the incorporation of the carborane-containing polymers in carcinoma cells. These findings pave the way for future investigations of these polymeric architectures as delivery agents for boron neutron capture therapy of cancer.

Introduction

There has been a growing interest in incorporating carboranes into tailored macromolecular structures by means of living or quasi-living polymerizations (*1–3*). This gain of interest stems from the rising awareness that carborane-based macromolecules can span a gamut of applications, for example: non-linear optical materials (*4–9*), precursors for ceramics (*10–13*), boron neutron capture therapy of cancer (BNCT) (*14*, *15*), *etc.* The work reported in this chapter is particularly relevant to the latter application. Adronov was the first to report the direct incorporation of carborane moieties in well-defined macromolecules (PDI < 1.2) by controlled radical polymerization, and by sequential dendrimer synthesis

(*2*, *16–18*). In their controlled radical experiment, they have demonstrated the possibility of incorporating carboranes in polymers of narrow polydispersity. However, loss of control upon attempted formation of the second block hampered the formation of well-defined amphiphilic structures (*2*, *16*). Seminal work by Sneddon and coworkers on precursors for boron-containing ceramics have highlighted the use of ring-opening metathesis polymerization (ROMP) to obtain high-boron content polymers that can undergo pyrolysis to form a boron carbide/carbon network (*1*, *19–22*). Building upon this work, Malenfant *et al.* have recently shown how ROMP can be used for the synthesis of boron-containing diblock copolymers and their subsequent transformation into nano-ordered ceramics (*3*).

In the following chapter, the synthesis of a boron-containing monomer based on an oxonorbornene functionalized *o*-carborane and its subsequent ring-opening metathesis polymerization to obtain low-dispersity (PDI < 1.1) high-molecular weight carborane-based polymers is described. The formation of amphiphilic block copolymers is demonstrated via sequential monomer addition, followed by deprotection of the second block to generate the amphiphilic copolymers. Initial solution behavior studies by dynamic light scattering (DLS) and atomic force microscopy (AFM) are also discussed. Once the diblock was characterized, post-polymerization modifications on the pendant amine are explored and subsequent cell studies are reported. Additionally, an alternative route using norbornene esters with polyethylene glycol (PEG) chains is also investigated. These copolymers are unique examples of a narrow polydispersity amphiphilic architectures incorporating carboranes that paves the way for the exploration of potential applications ranging from nanoscopic templates in thin film applications (*23*), to boron neutron capture therapy for the treatment of cancer (*14*, *24*).

Experimental Section

All organic reactants and reagents were purchased from Aldrich Chemical Co. and used as received unless specified otherwise. Anhydrous benzene and diethyl ether were dried over sodium with benzophenone as an indicator. Dichloromethane (DCM) was obtained through Fisher, distilled over calcium hydride and degassed with freeze-pump-thaw cycles. O-carborane was supplied by KatChem (*25*). Grubbs' third generation catalyst was synthesized according to the literature procedure and stored under nitrogen atmosphere (*26*).

Gel permeation chromatography (GPC) measurements for the homopolymers 4 were performed in tetrahydrofuran (THF) at 1.0 mL/min using a Knauer K-501 Pump with a K-2301 refractive index detector and a K-2600 UV detector, and a column bank consisting of two Polymer Labs PLGel Mixed D columns at 40°C. All other measurements were performed using a similar system with a column banks consisiting of three Polymer Labs PLGel Mixed D columns at 40°C. Molecular weights are reported relative to polystyrene standards.

^{1}H NMR, ^{11}B NMR and ^{13}C NMR were recorded at 300 MHz, 128 MHz and 100 MHz respectively on a Bruker NMR spectrometer at room temperature in deuterated chloroform. The individual NMR spectral assignments do not list the carborane B-H resonances. Due to the quadrupolar nature of boron, the resonances for the 10 B-H are observed as broad multiplets (δ (ppm) = 3.20-1.01). The integration of these multiplets accounts for 10 H. Melting points were measured with a melting point apparatus Electrochemical Mel-Temp in a capillary tube.

Dynamic Light Scattering (DLS) was performed in 18 mΩ water. The sample was prepared as followed; 30 mg of deprotected polymer were first dissolved in dimethyl sulfoxide DMSO (30mL). To this solution, 30 mL of 18 mΩ water were added using a syringe pump at a rate of 15 mL/h. The solution was then transferred into a dialysis bag and dialyzed against 2 L of 18 mΩ water for 2 days, renewing the water three times along the process. All experiments were performed at room temperature using an ALV unit equipped with an ALV/SP-125 precision goniometer (ALV-Laser Vertriebsgessellschaft m.b.h.), an Innova 70 argon laser (λ_0 = 514.5 nm, max. power 3 W, Coherent Inc.) operated at 300 mW, and a photomultiplier detector (Thorn EMI Electron Tubes). Signal from the detector was processed by an ALV 5000 Multiple Tau Digital Correlator board and associated software. The solutions were prefiltered through PVDF filters of 0.25-mm pore size. The experiment was performed using a polymer (PDI = 1.10, Mn = 30.500 g/mol with complete deprotection as monitored by ^{1}H NMR, molar ratio (SONIC/BONIA) = 2/1).

The investigation of the shape and topography of the micelles was performed using Atomic forceMicroscopy (Digital Instrument, Dimension TM 3100) using Nanoscope ® IIIa, version5.12r3 software, in tapping mode.

For the cell studies, monolayers of HEp-2 human laryngeal carcinoma cells were grown to approximately 80% confluency in 10% fetal bovine serum (FBS, Atlanta, Biologicals, Lawrenceville, GA) and imemzo (Richter's improved MEM insulin, Irvino Scientific, Santa, Ana CA). The media was then removed and 195 µL of the dialyzed solution of polymer was added to 250 µL 10% FBS/Cyclohexamide (Cyclohexamide overlay medium base, Cambrex, Walkersville, MD) and put on cells (*27*). The cells were incubated for about 22 h and were then washed three times with sterile phosphate buffer solution. The washed cells were then mounted with DAPI mounting medium. The viability of the cells was measured using a Trypan blue exclusion test.

Synthesis and Spectral Data

Preparation of the 1-(tert-Butyl-dimethylsilanyl)-1,2-dicarba-closo-dodecaborane(12) 2

Compound **2** was prepared using a slight modification of a Hawthorne procedure (*28*). In a round-bottom 500 mL Schlenk flask capped with a rubber septum and equipped with a magnetic stir bar. Under a nitrogen atmosphere, sublimed o-carborane **1** (15 g, 104 mmol, 1 eq) in a 2:1 mixture of benzene/ether (40 mL/20 mL) was deprotonated using n-butyl-lithium (78 mL, nBuLi 1.6 M, 125 mmol, 1.2 eq) at 0 °C dropwise. After 45 min, t-butyldimethylsilyl (TBDMS) chloride (TBDMS-Cl, 20.4 g, 135 mmol, 1.3 eq) in 30 mL of a benzene/ether (2:1 V:V) solution was added. After the addition of the silyl compound, the stopcock was closed and the septum removed. A rubber septum-capped condenser was fitted on top of the reaction flask with a nitrogen overpressure. The stopcock was then opened and the solution was brought to reflux. After refluxing overnight, the solution was quenched with a few drops of methanol to prevent exothermic reaction upon addition of 40 mL of water. The layers were separated in a separatory funnel and the aqueous solution was extracted with additional Et_2O (2 x 40 mL). The combined extracts were dried over $MgSO_4$ and concentrated *in vacuo* to give a slightly yellowish solid that was then distilled (85 °C, 20 mTorrs) to yield 24.7 g of a white powder, silyl-o-carborane **2** (92% yield). Also, there was no trace of remaining o-carborane. To verify the absence of any starting material in the final product, a careful sweep of temperatures, from 20°C up to 85 °C, at constant pressure (10 mTorrs) was performed with no apparent condensation of any solid prior to the final temperature. Conversely, keeping a constant temperature (85 °C) and carefully decreasing the pressure in the system, from 1 atm down to 10 mTorrs, no remaining o-carborane condensation was noted. High Resolution MS (EI+): m/z calcd 258.2758, found 258.4915; ^{1}H NMR ($CDCl_3$): δ (ppm) = 3.46 (s, 1H, CH), 1.03 (s, 9H, CCH_3), 0.26 (s, 6H, $SiCH_3$); ^{13}C NMR ($CDCl_3$): δ (ppm) = 66.02, 60.25, 26.98, 19.32, -4.56 ; ^{11}B {^{1}H} NMR ($CDCl_3$): δ (ppm) = 0.28 (1B), -1.80 (1B), -7.03 (2B), -10.75 (2B), -13.28 (2B); mp = 61°C.

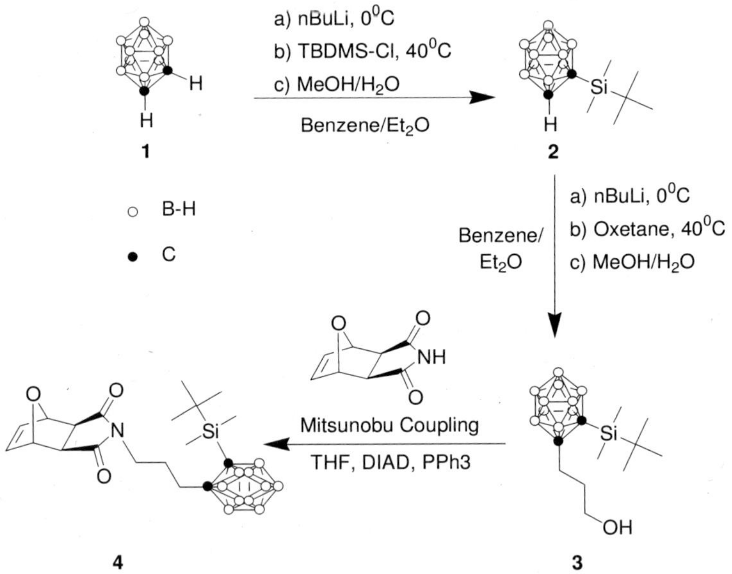

*Figure 1. Synthesis of the SONIC monomer **3**. TBDMS: terbutyldimethylsilyl. DIAD: diisopropylazodicarboxylate.*

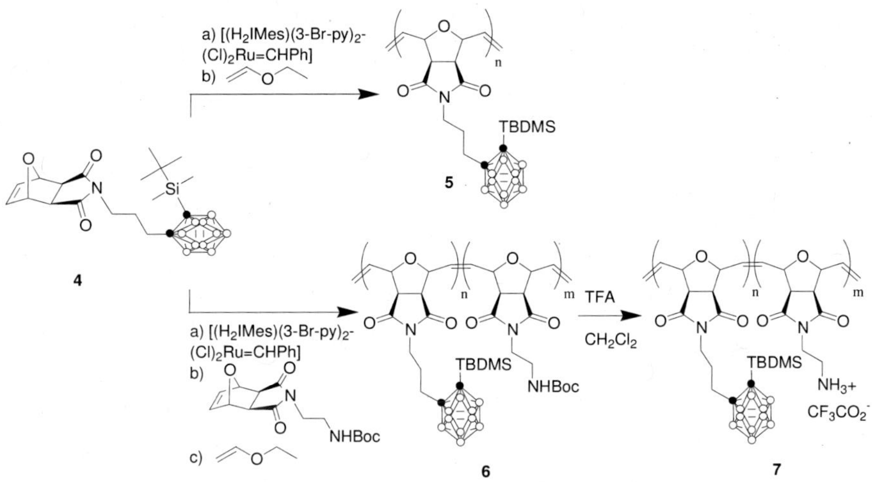

Figure 2. Ring-opening metathesis polymerization and subsequent amine deprotection syntheses. TBDMS: tert-Butyldimethylsilyl, TFA: Trifluoroacetic acid.

Preparation of the [2-(tert-Butyl-dimethyl-silanyl)-1,2-dicarba-closo-dodecaboran(12)-1-yl]-propan-1-ol 3

Compound **3** was prepared using a slight modification of a Hawthorne procedure in a setup comparable to the one used for **2** (*28*). Under a nitrogen atmosphere, compound 1 (15 g, 58 mmol, 1 eq) in benzene/ether (2:1 V:V) was deprotonated using dropwise addition of nBuLi (43.5 mL of a 1.6 M solution, 69.6 mmol, 1.2 eq) at 0°C. Trimethylene oxide (4.9 mL, 75.4 mmol, 1.3 eq) was added after 45 min and the solution turned light yellow. The solution was refluxed overnight and turned orange. The reaction was quenched with a few drops of methanol and 40 mL of water. The layers were separated in a separatory funnel and the aqueous solution was extracted with additional Et_2O (2 x 40 mL). The combined extracts were dried over $MgSO_4$ and concentrated *in vacuo* to give a sticky solid. The paste was then recrystallized from boiling hexanes followed by cooling at -4 °C. The thin fibrous white crystals were then filtered over sintered glass to obtain 15.6 g of carboranylpropanol **3** (85% yield). High Resolution MS (EI+): m/z calcd 315.3160, found 315.3155; ^{1}H NMR ($CDCl_3$): δ (ppm) = 3.69 (q, 2H, CH_2OH), 2.34 (qu, 2H, CH_2CH_2OH), 1.78 (t, 2H, CH_2-Cage), 1.06 (s, 9H, CCH_3), 0.31 (s, 6H, $SiCH_3$);); ^{13}C NMR ($CDCl_3$): δ (ppm) = 81.28, 76.47, 61.55, 35.51, 34.68, 27.80, 20.35, -2.48; ^{11}B {^{1}H} NMR ($CDCl_3$): δ (ppm) = 0.21 (1B), -3.95 (1B), -7.33 (2B), -10.31 (6B); mp = 83°C.

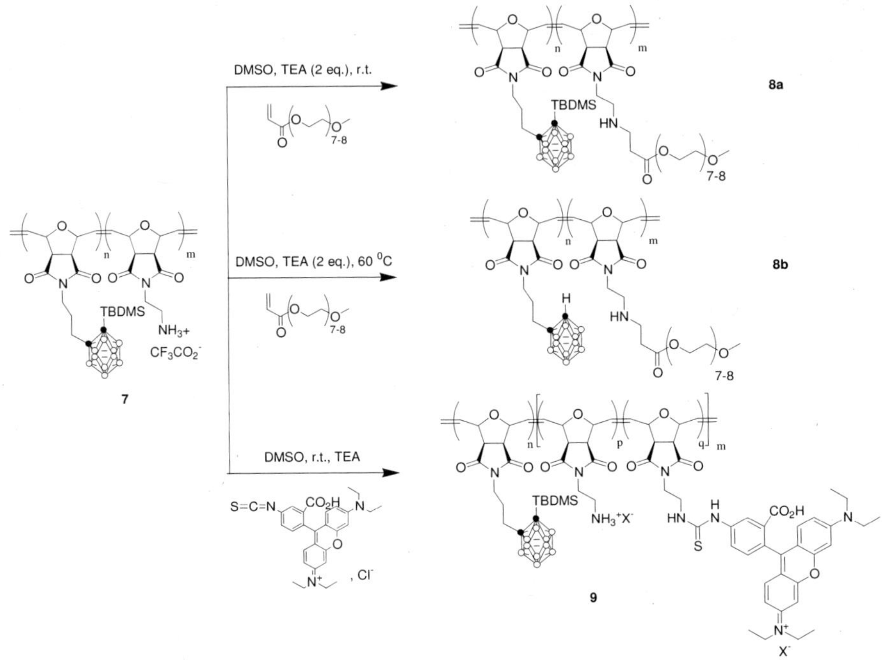

Figure 3. Post-polymerization modifications.

Preparation of (1R,2R,6S,7S)-4-{3-[2-(tert-Butyl-dimethylsilanyl)-1,2-dicarba-closo-dodecaboran(12)-1-yl]propyl}-10-oxa-4-aza-tricyclo[5.2.1.0^{2,6}]dec-8-ene-3,5-dione Referred to as Silyl-Protected OxoNorbornene Imide Carborane (SONIC) 4

In a dry 500 mL flask, exo-7-oxonorbornene imide (7.6 g, 46 mmol, 1 eq), [2-(tert-butyl-dimethyl-silanyl)-1,2-dicarba-closo-dodecaboran(12)-1-yl]-propan-1-ol **3** (16 g, 51 mmol, 1.1 eq) and triphenylphosphine (13.3 g, 51 mmol, 1.1 eq) were dissolved upon stirring in 300 mL of freshly distilled dry tetrahydrofuran (THF). The flask was kept under nitrogen pressure and cooled down to 0°C with an ice bath. After 5 minutes, diisopropylazodicarboxylate (DIAD) (3.2 mL, 16.26 mmol, 1.1 eq) was added and the solution turned yellow. The reaction was allowed to react overnight (16 h) affording a deep-yellow solution. The crude mixture was then concentrated in vacuo to obtain a sticky yellow paste. The paste was then redissolved in minimum amount of THF and precipitated by dropwise addition into hexanes (800 mL). The precipitate was a white powder with a slight reddish tint. The powder was then redissolved and precipitated a second time. The precipitate was then recrystallized from boiling methanol followed by cooling at -4°C, affording 18.6 g of flat square white crystals (87% yield). High Resolution MS (FAB-): m/z calcd 463.3505, found 463.3560; ^{1}H NMR ($CDCl_3$): δ (ppm) = 6.52 (s, 2H, CH=CH), 5.24 (s, 2H, bridgeheads), 3.46 (t, 2H, NCH_2), 2.85 (s, 2H, NC(O)CH), 2.09 (m, 2H, NCH_2CH_2), 1.82 (m, 2H, $NCH_2CH_2CH_2$), 1.06 (s, 9H, CCH_3) 0.32 (s, 6H,

$SiCH_3$); ^{13}C NMR ($CDCl_3$): δ (ppm) = 176.13, 136.53, 80.96, 80.37, 75.92, 47.43, 37.97, 34.98, 28.31, 27.51, 20.39, -2.60; ^{11}B {^1H} NMR ($CDCl_3$): δ (ppm) = 0.31 (1B), -3.85 (1B), -7.49 (2B), -10.25 (6B); mp = 136°C.

Homopolymerization of SONIC 5

The procedure was the same for every homopolymerization. Here is a standard procedure for the reactions carried out in septum vials, under constant agitation and under nitrogen atmosphere. To a 1 mL of solution of the desired amount of GIII, 200 mg of monomer **4** (4.31mmol) in 10 mL of dichloromethane were introduced rapidly and stirred for 6 minutes. After that time, the reaction was quenched with an excess of ethylvinylether and precipitated dropwise in cold methanol (150 mL) to afford a white powder. Table 1 shows the results obtained for different monomer to catalyst ratios. ^1H NMR ($CDCl_3$): δ (ppm) = 6.09 (br s, 0.84H, olefin trans), 5.82 (br s, 1.16H, olefin cis), 5.02 (br t, 1.16, CHO, cis), 4.46 (br s, 0.84H, CHO, trans), 3.6-3.2 (m, 4H, N-CH_2, CH-C(O)), 3.0-1.3 (br, 14H, BH, $\underline{CH_2}\underline{CH_2}$), 1.08 (s, 9H, CCH_3), 0.34 (s, 6H, $SiCH_3$); ^{13}C NMR ($CDCl_3$): δ (ppm) = 175.31, 131.10, 80.96, 80.18, 76.24, 52.37, 38.07, 35.13, 28.46, 27.54, 20.40, -2.47.

Diblock Copolymer Synthesis 6

In a screw cap vial under nitrogen atmosphere and constant agitation, the desired amount of SONIC was dissolved in 5mL of DCM and the catalyst was introduced as a 10 mM solution in DCM. After 5 min, {2-[(1R,2R,6S,7S)-3,5-Dioxo-10-oxa-4-aza-tricyclo[5.2.1.02,6]dec-8-en-4-yl]-ethyl}-carbamic acid *tert*-butyl ester referred to as Boc-protected OxoNorbornene Imide Amine (BONIA) (*29*), dissolved in 5 mL of DCM, was introduced into the reaction media using a syringe. The reaction was quenched after 10min by addition of excess ethyl vinylether. The solution was then precipitated into 50 mL of hexanes and filtered over sintered glass as a white solid.

Deprotection; Synthesis of an Amphiphilic Diblock Copolymer 7

In a 35 mL round-bottom flask, the polymer (150 mg) was dissolved in 4 mL of DCM and 15 mL of trifluoroacetic acid and allowed to react overnight. The polymer salt was then precipitated in diethyl ether and filtered over sintered glass.

Post-Polymerization PEGylation 8a

In a 50 mL round-bottom flask, polymer **7** (1 equiv.) was dissolved in dimethylsulfoxide (20 mL). Triethylamine (2 equiv.) were added and the reaction was allowed to stir for 20 min. PEG-acrylate (10 equiv.) was then added to the mixture and the mixture was allowed to react for 2 days. To obtain **8b**, the same procedure was utilized except that the reaction temperature was increased to 60 °C and some of the polymer solution was dialyzed against THF after reaction.

Table 1. Tabulated data for the homopolymerization of SONIC

	Target DP_n	*Target M_n*[a]	*M_n*[a,b]	*PDI*[a]	*Yield*
5i	54	25,000	22,000	1.07	83%
5ii	108	50,000	58,600	1.09	82%
5iii	162	75,000	79,100	1.05	87%
5iv	216	100,000	97,600	1.03	89%

[a] $g.mol^{-1}$. [b] As measured by GPC vs polystyrene standards.

In both cases, the yields were extremely low (below 10%) and just enough was recovered for ^1H NMR.

Attachment of Rhodamine-B 9

In a 25 mL round-bottom flask, polymer **7** (1g, 1 equiv.) was dissolved in DMSO (15 mL) and rhodamine B isothiocyanate (40 mg, 0.1 equiv.) was added to the mixture. A drop of triethylamine was then added and the mixture was allowed to react in the dark for 1 day. The mixture was then precipitated in cold diethyl ether to yield 850 mg of a pink powder (Yield 85%).

Results and Discussion

Gomez *et al.* had previously reported a protocol to synthesize the 3-[2-(*tert*-butyl-dimethylsilanyl)-1,2-dicarba-*closo*-dodecaboran(12)-yl)propan-1-ol **3** (Figure 1) (*28*). The synthesis of this entity was key to the elaboration of the targeted monomer **4**. However, while reproducing their experiments, slight modifications were introduced. The distillation of *o*-($C_2B_{10}H_{11}$)$SiMe_2$*t*Bu **2** occurred at higher pressure and lower temperatures than previously reported.

The oxanorbornene derivative **4** was synthesized via Mitsunobu coupling of exo-7-oxonorbornene imide and **3** (*30, 31*). A previous study in our group had demonstrated the difference in polymerization rates between exo and endo carborane-containing norbornene derivatives (*32*), corroborating findings by Grubbs and coworkers (*33*). In order to maximize the polymerization rates, an isomerically pure exo compound was needed. Furthermore, to avoid the problems of head-to head and head-to-tail additions, a symmetric monomer was desirable. To address these issues, we undertook the synthesis of SONIC (Silyl-protected OxaNorbornene Imide Carborane), whose symmetry and stereochemistry were specifically designed for faster and more controlled polymerization kinetics. This control permits the exploration of well-defined architectures, such as blocky structures.

Table 2. Tabulated data for the copolymerization of SONIC 4 and BONIA by ROMP

Entry	Molar Ratios SONIC	Molar Ratios BONIA	$DP_n(S)$	$DP_n(B)$	Target M_n^{a}	$M_n^{a,b}$	PDI^{b}
6i	4	5	22	28	18,835	17,300	1.09
6ii	2	1	50	25	30,894	32,990	1.10
6iii	8	5	54	34	35,523	33,800	1.12
6iv	4	5	54	68	46,006	45,500	1.15
6v	4	5	86	108	73,178	65,100	1.11

[a] $g.mol^{-1}$. [b] As measured by GPC vs polystyrene standards.

The homopolymerization of SONIC afforded polySONIC **5** in good yields (> 82%) and very narrow polydispersity (PDI < 1.1). To achieve this task, advantage was taken of the versatility and efficacy of ROMP catalysts, that not only give a living character to the polymerization, but also display a greater tolerance to diverse chemical functionalities (*26*). The reaction happens indeed within minutes and offers good control over molecular weight and molecular weight distribution, with no apparent broadening even for the highest M_n.

The diblock copolymers were obtained by sequential addition of SONIC and BONIA (Boc-protected OxoNorbornene Imide Amine) (Figure 2). These copolymers were obtained in good yields (> 80%) and with molecular weights ranging from 17 to 65 kg/mol. Also, the favorable comparison, between estimated and measured molecular weights, suggests that the polymers obtained behave like random coils in tetrahydrofuran. Even at high molecular weights, the polydispersity remains very narrow (Figure 4; Table 2, Entry 6v). The gel permeation chromatogram illustrates the increase in molecular weight upon addition of the second block (Figure 4). In addition, the molar feed and incorporation ratios are within 5% of one another as determined by ^{1}H-NMR spectroscopy.

Homopolymers of BONIA had previously been reported by our group as potential antibacterial polymers (*29*). Their low toxicity to red blood cells opens the prospects of medical applications for the synthesized diblock copolymers. The amphiphilicity, necessary for the formation of micelles that can be used as boron neutron capture therapy agents (*14*), was achieved through a post polymerization cleavage of the *t*Boc protecting group in the second block. In so doing, the copolymers obtained are a unique example of a well-defined (PDI < 1.15) amphiphilic copolymer containing carborane. The deprotection can be monitored by the disappearance of the characteristic N-H carbamate peak at δ = 6.89 ppm integrating for 1H and the disappearance of the 9H from the *tert*-butyl protons at δ = 1.35 ppm. At the same time, the appearance of the ammonium peak at δ = 8.01 ppm integrating for 3H is observed. As observed by ^{1}H NMR spectroscopy, the cleavage is quantitative.

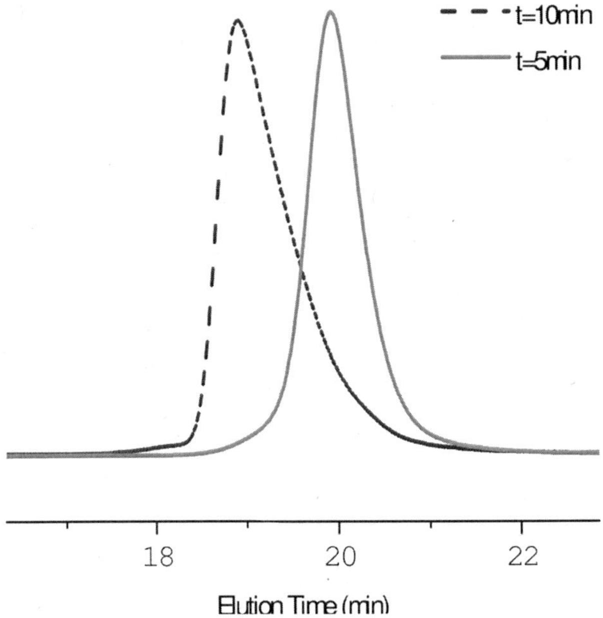

Figure 4. GPC chromatograms in THF of poly(SONIC) (solid line, PDI = 1.08, M_n = 37 kg/mol) and poly(SONIC-b-BONIA) (dashed line, Entry 6v in Table 2, PDI = 1.11 M_n = 65 kg/mol).

In addition, solution studies in water, using DLS, show the propensity of these polymers to aggregate. This phenomenon can be imputed to the formation of micelles were the charged hydrophilic block constitute the corona and the carborane containing block the core. These aggregates have an average hydrodynamic radius (R_h) of 41nm (Figure 6).

Atomic force microscopy was utilized to confirm the presence of aggregates and evaluate their shape. Figure 6 shows the presence of circular objects on a silica surface. These objects are found to have a radius of approximately 87 nm. This number is much larger than the hydrodynamic radius found in solution by DLS. However, the micelles are expected to flatten out as the positively charged ammoniums interact with the negatively charged silanols of the silica surface, leading to an apparent increase in size. This interaction is confirmed when looking at the profile of these cylindrical objects, as they are measured to be only 10 nm tall. A back of the envelope calculation allows us to calculate both the volume of a sphere with a 40 nm radius and a 10 nm-tall cylinder with a radius of 90 nm to be approximately 250 nm^3. Therefore, there seems to be a conservation of volume and these results lead us to believe that the value obtained by DLS is probably more accurate. Nevertheless, AFM provides some valuable insight into the shape of the micellar objects. Additionaly, the size might appear bigger because of the finite curvature of the AFM tip.

As illustrated by Figure 3, the deprotected amine provides us with a number of different post-polymerization possibilities. In particular, the PEGylation of the amine via a Michael addition of PEG-acrylate has been investigated at different

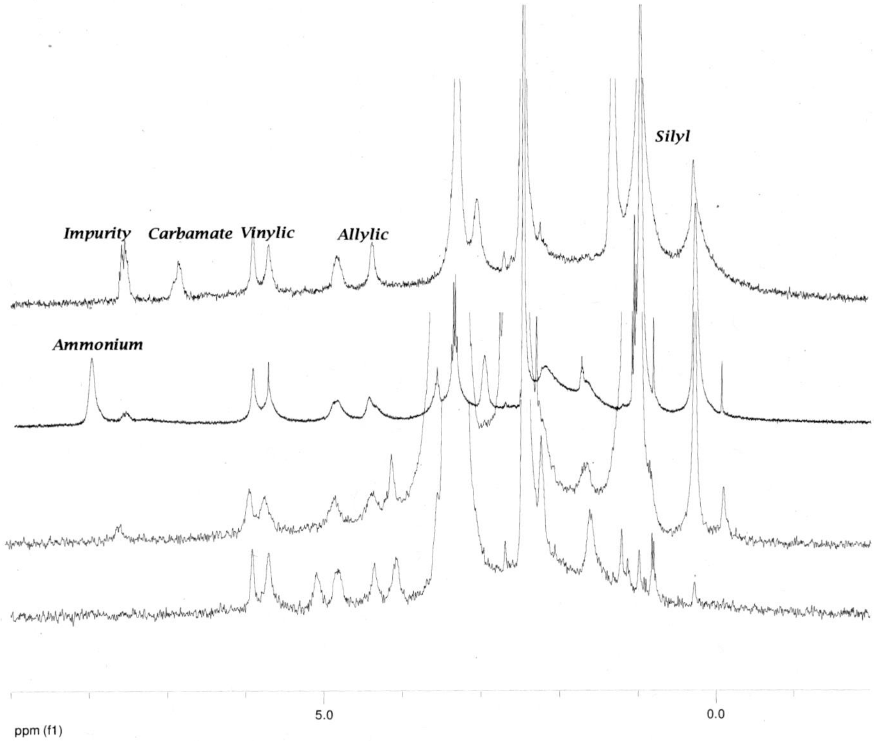

Figure 5. Illustration of the appearance of the ammonium peak at 8 ppm.

temperatures. In both cases, the disappearance of the ammonium peak at δ = 8.0 ppm was observed as well as the appearance of the characteristic PEG signal at 3.5 ppm, Figure 7. The synthesis of **8b** was attempted as a way to accelerate the kinetics of the reaction by increasing the reaction temperature. However, after taking an aliquot after 1 day, it was observed that the signals corresponding to the silyl group at δ = 1.06 and 0.33 ppm had disappeared, while a new signal at 5.1 ppm had appeared (Figure 5). The latter signal is the characteristic resonance of the C-H bond of carborane. Although the mechanism still remains unclear, it seems that the presence of triethylamine at high temperature is directly correlated with cleavage of the TBDMS group. A similar observation has been made when trying to hydrogenate polymer **5** in the presence of tri-*n*-propylamine, as will be discussed in chapter 3.

The introduction of PEG side chains is particularly important for the utilization of polymers in intravenous systemic delivery. Firstly, the PEG chains act as a hydrated cushion that prevents the adhesion of opsonizing proteins (*34*). By making the polymer "stealthy" to the immune system, the PEG chains increase the blood circulation half-time of the polymer which results in better accumulation at the desired location (*35*). Secondly, the bottle-brush architecture is believed to limit renal excretion by hampering the reptation of the macromolecules in the kidney (*36*). Again, this phenomenon also prolongs the circulation of the drug, rendering it more efficacious.

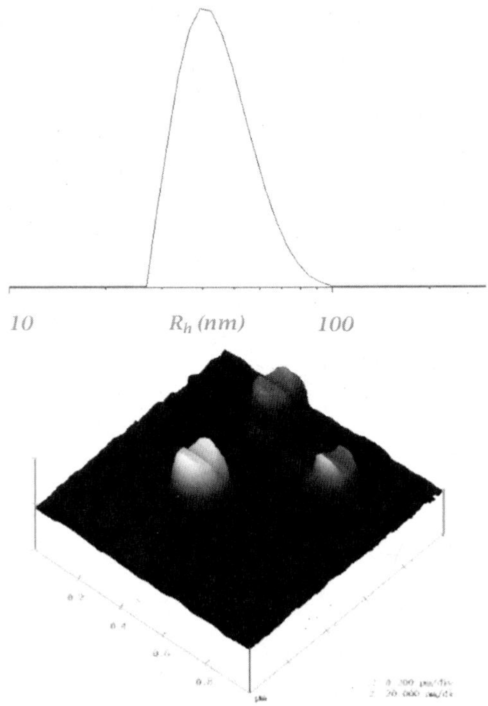

Figure 6. Distribution of sizes measured by dynamic light scattering (left) and atomic force microscope image (right) of micellar aggregates (Entry 6v of Table 2).

For topical applications, the presence of PEG chains is not quite as relevant as the risks of opsonization are greatly reduced. Therefore, it is very interesting to study the interaction of the charged polymer with carcinoma cells. In order to study this interplay *in vitro*, the attachment of a rhodamine B isocyanate was also undertaken (Figure 3). The yields were increased in comparison to those of the PEGylation reaction, as the solubility of the PEG groups complicates the precipitation step. Nevertheless, it was still difficult to quantify the level of attachment. In particular, the ionic nature of the polymer was a limiting factor for its injection in a GPC apparatus, to monitor the dye attachment to the polymer by UV detection. During the dialysis process of polymer **9** against water, a light pink coloration appeared outside of the dialysis bag. This phenomenon indicates that there is still some unattached dye that was not removed by the precipitation in diethyl ether. However, the coloration gradually faded upon continuous changing of the dialysis water. After 3 days, virtually no color change was observed. The presence of a dark pink color inside the bag suggests that the majority of the water-soluble dye had, in fact, coupled with the amine. The polymer was dialyzed for another 2.5 weeks to ensure that no residual dye would be present.

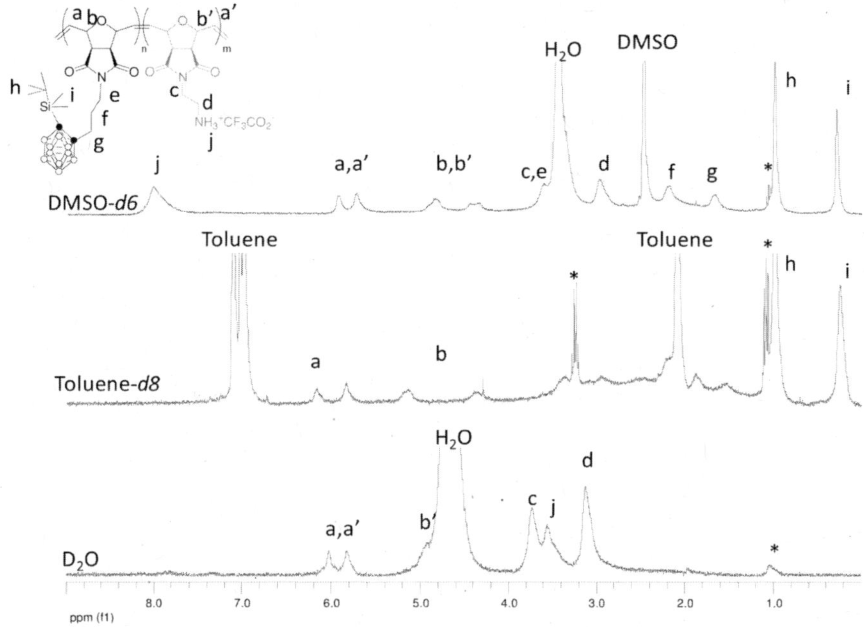

Figure 7. ^1H NMR experiment illustrating the solvents of different polarities.

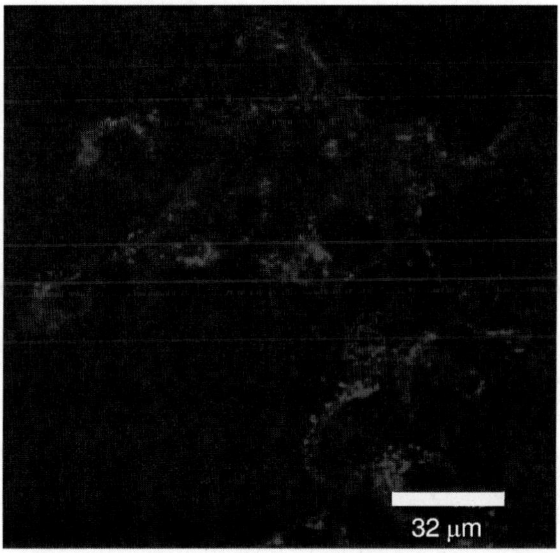

Figure 8. Confocal microscopy image of HEp-2 carcinoma cells after a 22h incubation period. (see color insert)

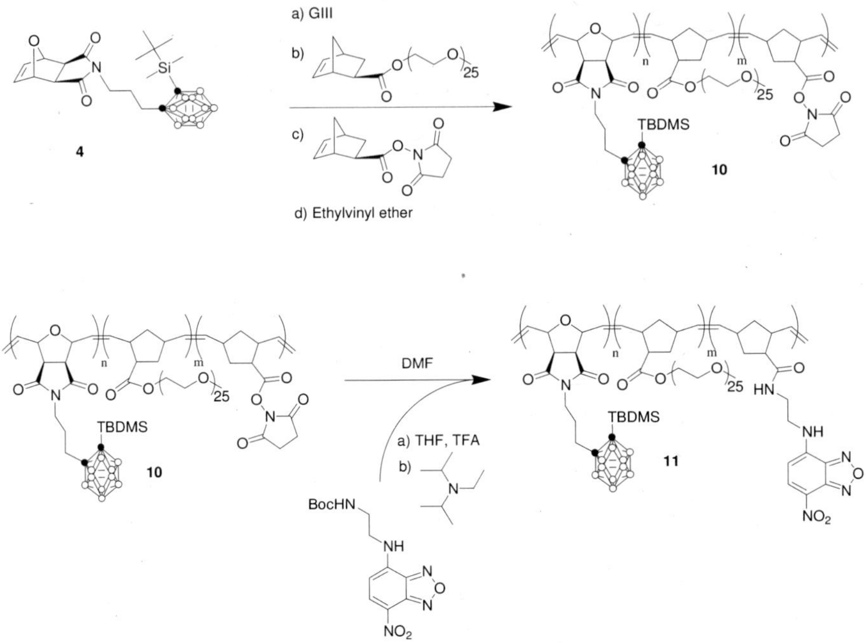

Figure 9. Triblock copolymer formation and labeling.

Confocal microscopy was utilized to evaluate the use of polymer **9** as potential agent for BNCT. Figure 8 illustrates the results obtained upon incubation of HEp-2 carcinoma cells with a solution of dialyzed polymer. Three very important observations have to be made. First of all, the internalization of labeled polymer **9** is apparent as illustrated by the z-sections taken on the cells. Secondly, the majority of the internalized polymer appears to be preferentially located around the nucleus of the cells. This observation is important as it has been cited that BNCT is more efficient when the delivery agent releases its cytotoxic payload closer to the nucleus of the cancerous cell. Finally, it is important to remark that without the presence of any neutron source, polymer **9** showed no cytotoxic effect after 22 h of incubation on cells in culture. The viability of cells was determined by a Trypan blue exclusion test. The principle of this test is that live, hardy cells possess an undamaged cell membrane that will exclude certain dyes such as Trypan blue. Upon adding the dye, the number of cells in the microscope field that had internalized the dye was comparable between the experimental and control groups (~ 3 to 5%).

While the synthesis of **8a**, **8b** and **9** showed the versatility of the chemistry that can be performed on the deprotected amine, it proved difficult to characterize the compounds. Therefore, the very nature of these post-polymerization routes complicates their implementation on a larger scale. Consequently, a possible alternative to the synthetic routes described above was investigated using a PEG-norbornene ester.

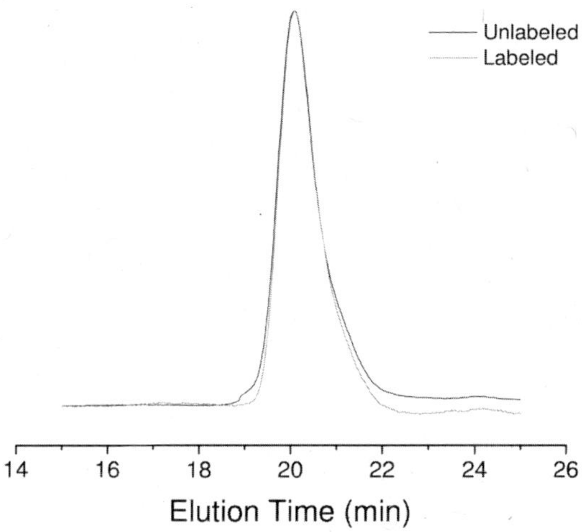

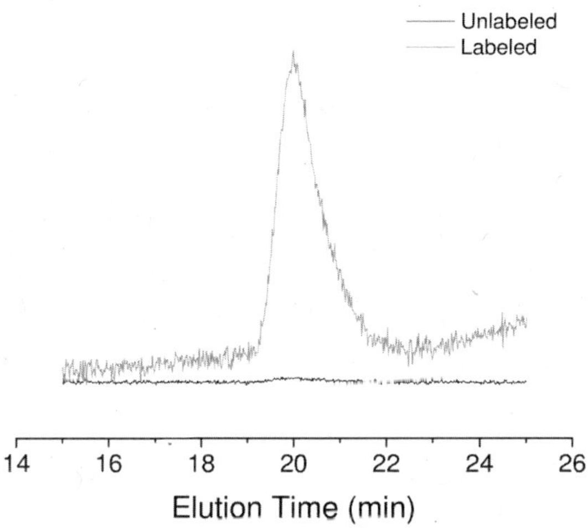

*Figure 10 RI (left) and UV (right) GPC chromatograms of **10** and **11**.*

Synthesis of Triblock Copolymers 10

In a 20 mL screw-cap vial, compound **4** (200 mg, 0.428 mmol) was dissolved in DMF (3 mL) and a 1 mL solution containing 10 mg of Grubbs third generation catalyst (0.011 mmol) was added. After 5 min, the PEG norbornene ester (200 mg, 0.167 mmol) in 3 mL of DMF was added and the mixture was allowed to react for 30 min. The NHS activated norbornene ester (6 mg, 0.22 mmol) was then introduced and the reaction was quenched after 3 min by addition of an excess of ethyl vinyl ether.The polymer was then precipitated in cold diethyl ether (cooled with liquid nitrogen) and filtered to yield 357 mg of a grey powder (88%).

Labelling of the Triblock Copolymer 11

In a 25 mL round-bottom flask, tBoc-protected nitrobenzoxydiaxole (NBD) (13 mg, 4 equiv.) was dissolved in THF (8 mL) and TFA was added (5 mL). After 4 h, the reaction mixture was syringed into another round-bottom flask containing polymer **10** (100 mg, 1 equiv.) in 10 mL of dry DMF. The reaction was allowed to stir overnight.

The alternative route described here presents the advantage of circumventing the post-polymerization PEGylation used in Route A by utilizing a PEG-functionalized monomer. While no solution studies have been performed on these materials, they are expected to lead to the formation of micellar assemblies comparable to that of Route A. The triblock copolymers were obtained by sequential additions of the monomers (Figure 9). The control obtained for the obtained polymer is remarkable as the PDIs are below 1.15. The number average molecular weight (22,500 g/mol) is however much inferior to the expected molecular weight (36,100 g/mol). This discrepancy can be accounted for by the fact that comb architecture display a lower apparent molecular weight by GPC (*37*). The attachment of the dye was monitored by GPC using UV detection. The GPC chromatograms (Figure 10) clearly show the detection of a refractive index change upon elution of the unlabeled polymer **10** and the NBD-labeled polymer **11**. However, by UV detection, the unlabelled polymer is not detected which indicates the attachment of the dye onto the polymer.

Conclusions

We have successfully developed a new monomer, SONIC, that enabled us to synthesize amphiphilic diblock copolymers containing carborane with low PDI (< 1.15) by two methods. The newly synthesized monomer was homopolymerized using ROMP yielding materials with low PDI (< 1.1). The solution behavior of the copolymers obtained by the polymerization-deprotection strategy (Route A) was characterized by DLS and AFM. It was shown that spherical micelles can be obtained. Post-polymerization modifications were successfully undertaken to attach PEG-acrylate via Michael addition and Rhodamine B isothiocyanate by nucleophilic substitituion. Additionally, preliminary cell studies showed the incorporation of the labeled-polymers inside carcinoma cells by confocal microscopy. An alternate method (Route B) was also investigated where the carborane-containing oxanorbornene was polymerized with a PEG-functionalized norbornene and an NHS activated norbornene ester. The NHS group allowed for facile labeling of the triblock copolymer as monitored by GPC. The obtained polymers by both routes are expected to provide a solid base for further exploration of amphiphilic carborane-based copolymers in BNCT applications.

References

1. Wei, X. L.; Carroll, P. J.; Sneddon, L. G. *Chem. Mater.* **2006**, *18*, 1113.
2. Parrott, M. C.; Marchington, E. B.; Valliant, J. F.; Adronov, A. *Macromol. Symp.* **2003**, *196*–201.

3. Malenfant, P. R. L.; Wan, J.; Taylor, S. T.; Manoharan, M. *Nat. Nanotechnol.* **2007**, *2*, 43.
4. Tsuboya, N.; Lamrani, M.; Hamasaki, R.; Ito, M.; Mitsuishi, M.; Miyashita, T.; Yamamoto, Y. *J. Mater. Chem.* **2002**, *12*, 2701.
5. Taylor, J.; Caruso, J.; Newlon, A.; Englich, U.; Ruhlandt-Senge, K.; Spencer, J. T. *Inorg. Chem.* **2001**, *40*, 3381.
6. Hamasaki, R.; Ito, M.; Lamrani, M.; Mitsuishi, M.; Miyashita, T.; Yamamoto, Y. *J. Mater. Chem.* **2003**, *13*, 21.
7. Entwistle, C. D.; Marder, T. B. *Angew. Chem., Int. Ed.* **2002**, *41*, 2927.
8. Allis, D. G.; Spencer, J. T. *Inorg. Chem.* **2001**, *40*, 3373.
9. Allis, D. G.; Spencer, J. T. *J. Organomet. Chem.* **2000**, *614*, 309.
10. Zalavutdinov, R. K.; Gorodetsky, A. E.; Zakharov, A. P. *Mikrochim. Acta* **1994**, *114-115*, 533.
11. Solozhenko, V. L.; Dubrovinskaia, N. A.; Dubrovinsky, L. S. *Appl. Phys. Lett.* **2004**, *85*, 1508.
12. Sneddon, L. G.; Su, K.; Fazen, P. J.; Lynch, A. T.; Remsen, E. E.; Beck, J. S. IUPAC Macromolecular Symposium Inorganic and Organometalic Oligomers and Polymers. *Pure Appl. Chem.* **1991**, *33*, 199.
13. Rybakov, S. Y.; Sharapov, V. M.; Gavrilov, L. E. *J. Phys. IV* **1995**, *5*, 921.
14. Hawthorne, M. F. *Angew. Chem., Int. Ed. Engl.* **1993**, 950.
15. Grimes, R. N. *J. Chem. Educ.* **2004**, *81*, 657.
16. Gratton, S. E. A.; Parrott, M. C.; Adronov, A. *J. Inorg. Organomet. Polym. Mater.* **2005**, *15*, 469.
17. Parrott, M. C.; Marchington, E. B.; Valliant, J. F.; Adronov, A. *J. Am. Chem. Soc.* **2005**, *127*, 12081.
18. Benhabbour, S. R.; Parrott, M. C.; Gratton, S. E. A.; Adronov, A. *Macromolecules* **2007**, *40*, 5678.
19. Wei, X. L.; Carroll, P. J.; Sneddon, L. G. *Organometallics* **2006**, *25*, 609.
20. Wei, X. L.; Carroll, P. J.; Sneddon, L. G. *Organometallics* **2004**, *23*, 163.
21. Pender, M. J.; Forsthoefel, K. M.; Sneddon, L. G. *Pure Appl. Chem.* **2003**, *75*, 1287.
22. Pender, M. J.; Carroll, P. J.Sneddon, L. G. *J. Am. Chem. Soc.* **2001**, *123*, 12222.
23. Thurn-Albrecht, T.; Steiner, R.; DeRouchey, J.; Stafford, C. M.; Huang, E.; Bal, M.; Tuominen, M.; Hawker, C. J.; Russell, T. *Adv. Mater. (Weinheim, Germany)* **2000**, *12*, 787.
24. Barth, R. F.; Coderre, J. A.; Vicente, M. G. H.; Blue, T. E. *Clin. Cancer Res.* **2005**, *11*, 3987.
25. Zheng, L.; Farris, R. J.; Coughlin, E. B. *J. Polym. Sci., Part A: Polym. Chem.* **2001**, *39*, 2920.
26. Love, J. A.; Morgan, J. P.; Trnka, T. M.; Grubbs, R. H. *Angew. Chem., Int. Ed.* **2002**, *41*, 4035.
27. Luo, L. B.; Tam, J.; Maysinger, D.; Eisenberg, A. *Bioconjugate Chem.* **2002**, *13*, 1259.
28. Gomez, F. A.; Johnson, S. E.; Hawthorne, M. F. *J. Am. Chem. Soc.* **1991**, *113*, 5915.

29. Ilker, M. F.; Nuesslein, K.; Tew, G. N.; Coughlin, E. B. *J. Am. Chem. Soc.* **2004**, *126*, 15870.
30. Messaoudi, S.; Anizon, F.; Pfeiffer, B.; Prudhomme, M. *Tetrahedron* **2005**, *61*, 7304.
31. Wang, Y.; Cai, J.; Rauscher, H.; Behm, R. J.; Goedel, W. A. *Chem.--Eur. J.* **2005**, *11*, 3968.
32. Simon, Y. C.; Coughlin, E. B. *Polym. Prepr. (Am. Chem. Soc., Div. Polym. Chem.)* **2005**, *46*, 771.
33. Grubbs, R. H. *Handbook of Metathesis*; Wiley, John & Sons, Incorporated: 2003, Vol. 3.
34. Owens, D. E.; Peppas, N. A. *Int. J. Pharm.* **2006**, *307*, 93.
35. Duncan, R. *Nat. Rev. Drug Discovery* **2003**, *2*, 347.
36. Gillies, E. R.; Frechet, J. M. J. *J. Am. Chem. Soc.* **2002**, *124*, 14137.
37. Nakamura, Y.; Wan, Y.; Mays, J. W.; Iatrou, H.; Hadjichristidis, N. *Macromolecules* **2000**, *33*, 8323.

Chapter 6

Synthesis and Characterization of Helix-Coil Block Copoly(α-peptoid)s

Li Guo and Donghui Zhang*

Department of Chemistry, Macromolecular Studies Group, Louisiana State University, Baton Rouge, LA 70803
***dhzhang@lsu.edu**

Helix-coil block copoly(α-peptoid)s [i.e., poly(*S*-*N*-CHMePh-glycine)-*b*-poly(*N*-Bu-glycine) (PNEPG-*b*-PNBG)] have been synthesized by *N*-heterocyclic carbene (NHC)-mediated sequential ring-opening polymerizations (ROP) of the corresponding *N*-substituted *N*-carboxyanhydride monomers (^{N}CHMePh-NCA and ^{N}Bu-NCA). The helical block degree of polymerization (DP) has been kept constant, whereas the coil block DP has been varied by changes to the ratio of the second monomer to initiator (i.e., [^{N}Bu-NCA]0:[NHC]$_0$). SEC analysis reveals an increase of polymer molecular weight (MW) as the [^{N}Bu-NCA]$_0$:[NHC]$_0$ ratio is increased, confirming the formation of block copoly(α-peptoid)s having reasonably narrow molecular weight distribution (PDI =1.15-1.26). The copolymer structure has been verified by ^{1}H and ^{13}C{^{1}H} NMR spectroscopy. Polymer MWs obtained by ^{1}H NMR analysis agree with the theoretical MWs for living polymerization. Circular dichroism (CD) analysis reveals that all block copoly(α-peptoid)s exhibit enhanced helical conformational stability than the PNEPG homopolymers, as manifested by the larger mean residual ellipticity (MRE) in the former. The degree of enhancement depends on the copolymer composition.

Introduction

Rod-coil block copolymers have attracted much attention due to their ability to self-assemble into unique morphologies with hierarchical structures both in solution and the solid state (*1–5*). In addition to the parameters that govern the self-assembly of coil-coil block copolymers (e.g., copolymer composition, chemistry, architecture, solvent, temperature, etc.), the aggregation tendency of the rigid rod segments is another important factor that impacts the self-assembly of rod-coil block copolymers (*1*, *2*). Helical poly(α-peptide)s have been examined as the rigid rod component in many rod-coil block copolymers (*1*). These helical rod-coil block copolymers are nanostructured materials with potential utility in many areas, e.g. biomedical science owing to the biocompatibility and biodegradability properties of poly(α-peptide)s as well as the stimuli responsive behavior resulting from the helix-to-coil transition (*6–10*).

Poly(α-peptoid)s are structural mimics of poly(α-peptide)s having polyglycine backbones and proteogenic or synthetic side-chains directly attached to the nitrogen. Due to the lack of hydrogen bonding interactions, the secondary structures of poly(α-peptoid)s are mainly governed by the backbone rigidity and steric characteristics of side chains (*11*, *12*). Additionally, poly(α-peptoid)s are more resistant against proteolysis and hydrolysis than poly(α-peptide)s and have emerged as a class of important peptidomimetic materials for applications including targeted drug delivery (*13*) and as antimicrobial agents (*14*, *15*). Previous applications have primarily focused on exploiting the molecular characteristics of oligomeric α-peptoids or poly(α-peptoid)s. Synthesis of block copoly(α-peptoid)s and investigation of their self-assembly represents a major step towards hierarchically structured peptidomimetic materials with variable length scale and dynamic range, potentially leading to their broader utility in the biomedical area.

Conventional synthetic methods towards poly(α-peptoid)s suffer low efficiency (*16–19*) and limited substrate scope (*20–24*). We recently reported an *N*-heterocyclic carbene (NHC)-mediated ring-opening polymerization (ROP) of *N*-substituted *N*-carboxyanhydride monomers (^{N}R-NCA) to afford cyclic poly(α-peptoid)s (*25*). The reaction exhibits the characteristics of a living polymerization, making it possible to synthesize block copoly(α-peptoid)s by sequential monomer addition. It has also been demonstrated that poly(α-peptoid)s bearing bulky chiral side-chains [e.g., (S)-CHMePh] adopt helical conformations reminiscent of polyproline I helices (*26*). In this contribution we investigate the synthesis of a series of helix-coil block copoly(α-peptoid)s [i.e., poly(*S*-*N*-CHMePh-glycine)-*b*-poly(*N*-Bu-glycine) (PNEPG-*b*-PNBG)] by NHC-mediated sequential ring-opening polymerizations of the corresponding ^{N}R-NCAs. The block copolymers have been characterized by ^{1}H and ^{13}C{^{1}H} NMR spectroscopy, size exclusion chromatography (SEC) and circular dichroism (CD) spectroscopy.

Experimental Section

General Considerations

All air- or moisture-sensitive compounds were handled under a nitrogen atmosphere, either using standard Schlenk-line techniques or in a glovebox. Anhydrous tetrahydrofuran (THF), hexane, dichloromethane and toluene

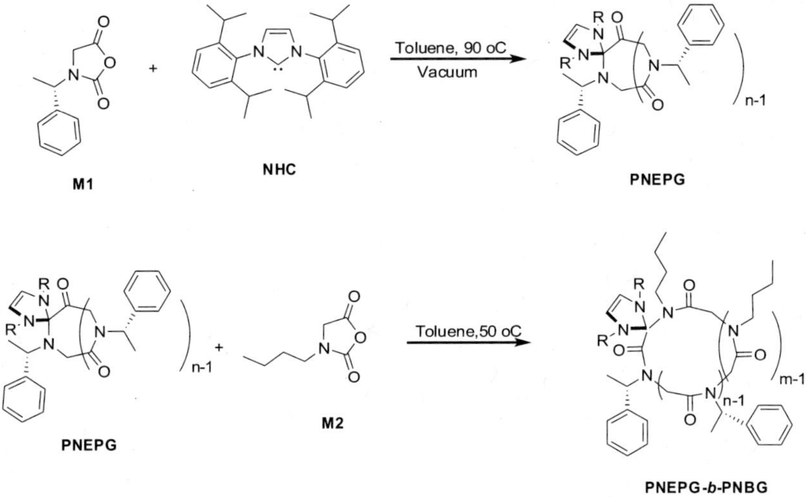

Scheme 1

were obtained by passing through activated alumina columns or molecular sieves under an argon atmosphere (Innovative Technology, Inc.). 2,6-Diisopropylphenylimidazol-2-ylidene (NHC) (*27*), *N*-(*S*)-CHMePh-NCA (**M1**) (*27*) and ^{N}Bu-NCA (**M2**) (*26*) have been synthesized by reported procedures.

^{1}H and ^{13}C{^{1}H} NMR spectra were recorded on a Bruker AV-400 spectrometer. Chemical shifts are reported in ppm and referenced to the solvent protio impurities and solvent ^{13}C{^{1}H} resonances. Size exclusion chromatography (SEC) was performed using an Agilent 1200 system (Agilent 1200 series degasser, isocratic pump, auto sampler and column heater) equipped with three Phenomenex 5 μm, 300×7.8 mm columns [100 Å, 1000 Å and Linear (2)], Wyatt DAWN EOS multi-angle light scattering (MALS) detector (GaAs 30 mW laser at λ=690 nm) and Wyatt Optilab rEX differential refractive index (DRI) detector with a 690 nm light source. DMF containing 0.1 M LiBr was used as the eluent at a flow rate of 0.5 mL·min^{-1}. The column temperature was 50 °C and the detector temperature was 25 °C for both MALS and DRI detectors.

Circular dichroism (CD) data were collected with the high tension voltage (i.e., the voltage applied to the photomultiplier) less than 600 V on a Jasco J810 CD spectrometer (Japan Spectroscopic Corporation) with a path length of 0.1 cm and a band width of 1 nm. Two scans were conducted and averaged between 185 nm and 250 nm at a scanning rate of 20 nm·min^{-1} with a resolution of 0.5 nm. The data were processed by subtracting the solvent background and smoothing with Means-Movement method with a convolution of 5. The CD spectra were reported in mean residue ellipticity (MRE) (unit: deg·cm^{2}·dmol^{-1}) which was calculated by the equation $[\theta]_\lambda = \mathrm{MRW} \times \theta_\lambda/[10 \times d \times c \times \mathrm{MW}_{PNEPG}/\mathrm{MW}_{PNEPG\text{-}b\text{-}PNBG}]$ (*28*), where MRW is the mean residue weight of the helix-forming segment (i.e., 161.2), θ_λ is the observed ellipticity (deg) at the wavelength λ, d is the path length, c is the concentration (g·mL^{-1}), MW_{PNEPG} and $\mathrm{MW}_{PNEPG\text{-}b\text{-}PNBG}$ are the MW of the PNEPG block and the respective PNEPG-*b*-PNBG block copolymers. Solutions

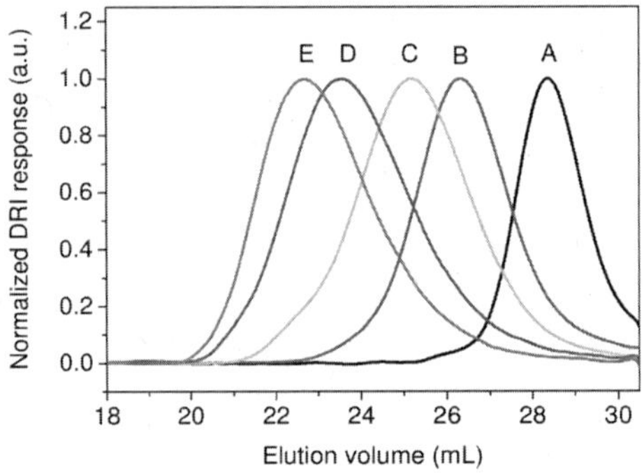

Figure 1. SEC chromatographs of poly(S-N-CHMePh-glycine) homopolymer and poly(S-N-CHMePh-glycine)-b-poly(N-Bu-glycine) block copolymer (A: $PNEPG_{23}$; B: $PNEPG_{23}$-b-$PNBG_{23}$; C: $PNEPG_{23}$-b-$PNBG_{42}$; D: $PNEPG_{23}$-b-$PNBG_{72}$; E: $PNEPG_{23}$-b-$PNBG_{89}$).

of copoly(α-peptoid)s in CH_3CN (0.02 mg·mL^{-1}) were stored for at least 48 hrs before measurement.

Representative Procedure for NHC-Mediated Sequential Polymerizations of (*S*)-^{N}CHMePh-NCA and ^{N}Bu-NCA

Inside a glovebox, *N*-(*S*)-CHMePh-NCA (**M1**) (138 mg, 0.67 mmol) was dissolved in toluene (1.5 mL) and a predetermined amount of 2,6-diisopropylphenylimidazol-2-ylidene (**NHC**)/toluene stock solution (193 μL, 1.3×10^{-2} mmol, 69.6 mM) was added by syringe. The reaction mixture was degassed by three freeze-pump-thaw cycles and then stirred at 90°C for 39 h under vacuum. An aliquot of the reaction solution (0.14 mmol, 350 μL) was added into the ^{N}Bu-NCA (**M2**) solution in toluene (350 μL, 0.14 mmol, 0.4 M) and stirred at 50°C for 49 h under a nitrogen atmosphere. The polymers were isolated by precipitation with hexane and filtration. Further purification by Soxhlet extraction with warm hexane and drying under vacuum afforded a yellow powder (16 mg, 62 % yield).

Results and Discussion

Block Copoly(α-peptoid)s Synthesis and Characterization

We have previously demonstrated that poly(*S*-*N*-CHMePh-glycine) (PNEPG) homopolymers adopt helical conformations reminiscent of polyproline I helices (*26*), whereas poly(*N*-Bu-glycine) (PNBG) adopt random-coil conformations (*25*). In this work we investigate the synthesis of helix-coil block copoly(α-peptoid)s containing a constant PNEPG block and a PNBG block with variable segment length (PNEPG-*b*-PNBG) by NHC-mediated sequential polymerizations of (*S*)-^{N}CHMePh-NCA (**M1**) and ^{N}Bu-NCA (**M2**) (Scheme 1). Polymerizations of **M1** and **M2** were allowed to reach nearly complete conversions. The block copolymers were isolated by precipitation with hexane and purified by Soxhlet extraction with warm hexane. The copolymer structure has been characterized by ^1H and $^{13}C\{^1H\}$ NMR spectroscopy. The absolute MW and PDI of the PNEPG homopolymer have been determined by size-exclusion chromatography coupled with multi-angle light scattering and differential refractive index detectors (SEC-MALS-DRI) using a measured dn/dc of 0.1396 (19) mL·g^{-1}. The experimental MW of the PNEPG block (M_n=3.7 kg·mol^{-1}) is lower than the theoretical prediction based on the [**M1**]$_0$:[**NHC**]$_0$ ratio and conversion (Table 1). This is due to intramolecular transamidation that competes with chain propagation, resulting in reduced polymer MWs and cyclic oligomer formation (*26*). The block copolymer MWs and the PDIs have been determined by SEC using a calibration curve based on polystyrene standards as well as by ^1H NMR analysis. All block copolymers (Figure 1) exhibit single monomodal SEC chromatograms with reasonably narrow molecular weight distribution (PDI=1.15-1.26). No low MW polymers or unreacted PNEPG were detected in the SEC chromatograms, consistent with a quantitative incorporation of PNEPG in the chain extension. Moreover, the SEC peaks are systematically shifted towards lower elution volumes for the block copolymers prepared from reactions with higher [**M2**]$_0$:[**NHC**]$_0$ ratio, suggesting a steady increase of the polymer MW.

Table 1. NHC-initiated polymerizations of (*S*)- ^{N}CHMePh-NCA (M1) with ^{N}Bu-NCA (M2)

Polymers[a]	*[M1]$_0$:[M2]$_0$:[NHC]$_0$*	*M_n (kg·mol^{-1})*			*PDI*
		Calcd[b]	*NMR*	*SEC*[d]	*SEC*[d]
$PNEPG_{23}$	50:0:1	3.6[c]	-	3.7[e]	1.07[e]
$PNEPG_{23}$-*b*-$PNBG_{23}$	50:25:1	6.5	6.3	11.6	1.15
$PNEPG_{23}$-*b*-$PNBG_{42}$	50:50:1	9.4	8.5	15.8	1.20
$PNEPG_{23}$-*b*-$PNBG_{72}$	50:100:1	15.0	11.8	22.5	1.26
$PNEPG_{23}$-*b*-$PNBG_{89}$	50:150:1	20.7	13.8	28.7	1.25

[a] the numbers in subscripts indicates the degree of polymerization (DP) calculated by SEC and ^1H NMR spectroscopy;
[b] theoretical MWs were calculated from the absolute MW of PNEPG block (M_n=3.7 kg·mol^{-1}) and the [**M2**]$_0$/[**NHC**]$_0$ ratio;
[c] MW was calculated from the percentage polymer content in the product and the [**M1**]$_0$/[**NHC**]$_0$ ratio (50:1);
[d] MW and PDI were obtained from a tandem SEC-MALS-DRI system using a calibration curve constructed with polystyrene standards;
[e] MW and PDI was obtained from a tandem SEC-MALS-DRI system using a measured dn/dc of 0.1396 (19) mL·g^{-1}.

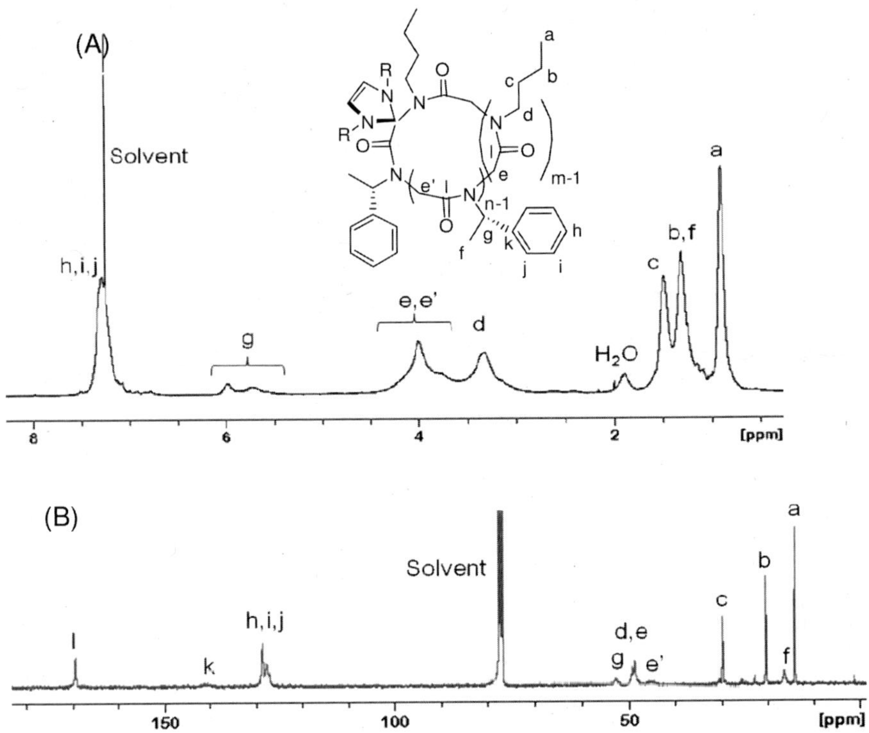

Figure 2. (A) ^{1}H and (B) $^{13}C\{^{1}H\}$ NMR spectra of PNEPG$_{23}$-b-PNBG$_{42}$ in $CDCl_3$.

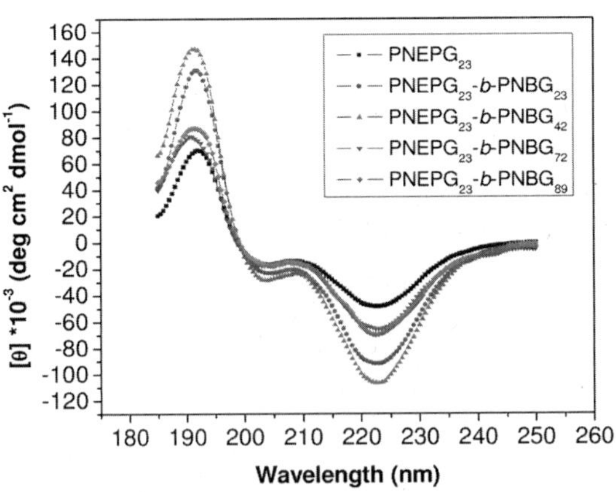

Figure 3. CD spectra of PNEPG homopolymer and PNEPG-b-PNBG block copolymer at 20 °C in CH_3CN at 0.02 mg·mL^{-1} concentration.

^{1}H and ^{13}C{^{1}H} NMR spectroscopy (Figure 2) has confirmed the block copoly(α-peptoid) structure (Scheme 1). Block copolymer compositions have been determined by integrating the methine protons (g, Figure 2A) on the PNEPG side-chains at 5.72-5.98 ppm relative to the methyl protons (a, Figure 2A) on the PNBG side-chains at 0.91 ppm. The absolute MWs of the block copolymers have been independently determined from the block copolymer composition and the absolute MW of the PNEPG block (Table 1). The absolute MW of the block copolymers are lower than theoretical predictions based on quantitative chain extension of the PNEPG blocks. The discrepancy is more pronounced in the high MW range, suggesting the occurance of competing side-reactions. While we observed minimal or no intramolecular transamidation in the NHC-mediated homopolymerization of **M2** (*25*), the PNEPG block may impose conformational changes on the PNBG segment, resulting in enhanced intramolecular transamidation and henced reduced MW.

CD Analysis

CD spectra of the PNEPG-*b*-PNBG block copolymers have been measured in acetonitrile at 20°C and reported as the mean residual ellipticity (MRE) normalized against the helix-forming block length. All block copolymers exhibit similar CD profiles with a positive ellipticity maximum at 192 nm and two negative ellipticity minima at 204 and 223 nm (Figure 3), consistent with the polyproline I helical conformations previously observed for PNEPG homopolymers (*26*) and *S*-*N*-CHMePh-glycine oligomers (*16*). All helix-coil block copolymers exhibit larger MREs than PNEPG homopolymers with identical helical chain length, suggesting enhanced conformational stability for the helices of the former. The enhancement is most pronounced for the $PNEPG_{23}$-*b*-$PNBG_{42}$ with an intermediate coil-block length. While it has been shown that longer PNEPG homopolymers adopt more stable helical conformations than their shorter analogs (attributed to folding cooperativity), it is not apparent why attachement of a coil segment enhances the helical stability of the PNEPG block. We are currently investigating this issue in greater detail.

Conclusions

Helix-coil block copoly(α-peptoid)s (i.e., PNEPG-*b*-PNBG) with different compositions have been successfully synthesized by NHC-mediated sequential polymerization of (*S*)-^{N}CHMePh-NCA and ^{N}Bu-NCA. The block copolymer structure has been characterized by ^{1}H and ^{13}C{^{1}H} NMR spectroscopy. The polymer MW and PDI have been determined by SEC and ^{1}H NMR analysis. CD analysis reveals that the PNEPG blocks maintain PPI helical conformations, which become stabilized by increasing coil-segment length (i.e., PNBG).

Acknowledgments

This work is supported by LSU, Louisiana Board of Regents [LEQSF(2008-11)-RD-A-11] and NSF (CHE 0955820).

References

1. Klok, H.-A.; Lecommandoux, S. *Adv. Mater.* **2001**, *13*, 1217–1229.
2. Olsen, B. D.; Segalman, R. A. *Mater. Sci. Eng. R Rep.* **2008**, *R62*, 37–66.
3. Jenekhe, S. A.; Chen, X. L. *Science* **1998**, *279*, 1903–1907.
4. Heiser, T.; Adamopoulos, G.; Brinkmann, M.; Giovanella, U.; Ould-Saad, S.; Brochon, C.; Wetering, K.; Hadziioannou, G. *Thin Solid Films* **2006**, *511-512*, 219–223.
5. Yan, L.; Tao, W. *J. Polym. Sci. Part A: Polym. Chem.* **2008**, *46*, 12–20.
6. Klok, H.-A. *PMSE Prepr.* **2004**, *90*, 245.
7. Zhang, J.; Yu, Z.; Wan, X.; Chen, X.; Zhou, Q. *Macromol. Rapid. Commun.* **2005**, *26*, 1241–1245.
8. Kuo, S.-W.; Lee, H.-F.; Huang, C.-F.; Huang, C.-J.; Chang, F.-C. *J. Polym. Sci., Part A: Polym. Chem.* **2008**, *46*, 3108–3119.
9. Sánchez-Ferrer, A.; Mezzenga, R. *Macromolecules* **2010**, *43*, 1093–1100.
10. Rao, J.; Luo, Z.; Ge, Z.; Liu, H.; Liu, S. *Biomacromolecules* **2007**, *8*, 3871–3878.
11. Baldauf, C.; Günther, R.; Hofmann, H.-J. *Phys. Biol.* **2006**, *3*, S1–S9.
12. Gorske, B. C.; Stringer, J. R.; Bastian, B. L.; Fowler, S. A.; Blackwell, H. E. *J. Am. Chem. Soc.* **2009**, *131*, 16555–16567.
13. Wender, P. A.; Mitchell, D. J.; Pattabiraman, K.; Pelkey, E. T.; Steinman, L.; Rothbard, J. B. *Proc. Natl. Acad. Sci.* **2000**, *97*, 13003–13008.
14. Patch, J. A.; Barron, A. E. *J. Am. Chem. Soc.* **2003**, *125*, 12092–12093.
15. Chongsiriwatana, N. P.; Patch, J. A.; Czyzewski, A. M.; Dohm, M. T.; Ivankin, A.; Gidalevita, D.; Zuckermann, R. N.; Barron, A. E. *Proc. Natl. Acad. Sci.* **2008**, *105*, 2794–2799.
16. Wu, C. W.; Kirshenbaum, K.; Sanborn, T. J.; Patch, J. A.; Huang, K.; Dill, K. A.; Zuckermann, R. N.; Barron, A. E. *J. Am. Chem. Soc.* **2003**, *125*, 13525–13530.
17. Patch, J. A.; Barron, A. E. *J. Am. Chem. Soc.* **2003**, *125*, 12092–12093.
18. Hara, T.; Durell, S. R.; Myers, M. C.; Appella, D. H. *J. Am. Chem. Soc.* **2006**, *128*, 1995–2004.
19. Zuckermann, R.N.; Kodadek, T. *Curr. Opin. Mol. Ther.* **2009**, *11*, 299–307.
20. Kricheldorf, H. R.; von Lossow, C.; Lomadze, N.; Schwarz, G. *J. Polym. Sci., Part A: Polym. Chem.* **2008**, *46*, 4012–4020.
21. Kricheldorf, H. R.; von Lossow, C.; Schwarz, G. *Macromolecules* **2005**, *38*, 5513–5518.
22. Kricheldorf, H. R.; von Lossow, C.; Schwarz, G. *J. Polym. Sci., Part A: Polym. Chem.* **2006**, *44*, 4680–4695.
23. Kricheldorf, H. R.; von Lossow, C.; Schwarz, G.; Fritsch, D. *Macromol. Chem. Phys.* **2005**, *206*, 1165–1170.

24. Kricheldorf, H. R.; von Lossow, C.; Schwarz, G. *Macromol. Chem. Phys.* **2004**, *205*, 918–924.
25. Guo, L.; Zhang, D. *J. Am. Chem. Soc.* **2009**, *131*, 18072–18074.
26. Guo, L.; Li, J.; Brown, Z.; Ghale, K.; Zhang, D. *Biopolym. Pept. Sci.*, in print.
27. Arduengo, A. J.; Krafczyk, R.; Schmutzler, R.; Craig, H. A.; Goerlich, J. R.; Marshall, W. J.; Unverzagt, M. *Tetrahedron* **1999**, *55*, 14523–14534.
28. Kelly, S. M.; Jess, T. J.; Price, N. C. *Biochim. Biophys. Acta* **2005**, *1751*, 119–139.

Chapter 7

Block Copolymer Synthesis through the Use of Switchable RAFT Agents

Graeme Moad,[1,*] Massimo Benaglia,[2] Ming Chen,[1] John Chiefari,[1] Yen K. Chong,[1] Daniel J. Keddie,[1] Ezio Rizzardo,[1] and San H. Thang[1]

[1]CSIRO Materials Science and Engineering, Bag 10, Clayton South, Victoria 3169, Australia
[2]ISOF-CNR, Area della Ricerca, Via P. Gobetti 101, 40129 Bologna, Italy
[*]graeme.moad@csiro.au

The use of "switchable" RAFT agents in expanding the use of RAFT polymerization in block copolymer synthesis is decsribed. *N*-(4-pyridinyl)-*N*-methyldithiocarbamates provide excellent control over polymerization of less activated monomers (LAMs) (e.g., vinyl acetate, *N*-vinylcarbazole), and, after addition of one equivalent of a protic or Lewis acid, become effective in controlling polymerization of more activated monomers (MAMs) (e.g., styrene, methyl methacrylate). They thus allow a convenient synthesis of polyMAM-*block*-polyLAM with narrow molecular weight distributions. Advantages, limitations and factors effecting the scope of the process are discussed.

Introduction

Reversible addition-fragmentation chain transfer (RAFT) polymerization is a reversible deactivation radical polymerization (*1*) and one of the most versatile processes for conferring living characteristics on radical polymerization (*2–6*). The polymerization of most radically polymerizable monomers can be controlled by the RAFT process. However, the RAFT agent (ZC(=S)S-R) needs to be selected according to the monomer(s) used. RAFT agents such as dithioesters (Z=aryl or alkyl) or trithiocarbonates (Z=alkylthio), suitable for controlling polymerization of the 'more-activated' monomers (MAMs) (e.g. MA, St) (*7*, *8*), inhibit or retard polymerizations of the 'less activated' monomers (LAMs) (e.g.

VAc, NVP). Similarly RAFT agents suitable for controlling polymerizations of LAMs, such as *N,N*-dialkyl- or *N*-alkyl-*N*-aryl dithiocarbamates and xanthates, tend to be ineffective with MAMs. As a consequence, the synthesis of narrow dispersity polyMAM-*block*-polyLAM is difficult or not possible using the conventional RAFT process.

Several groups (*9–17*) have reported strategies to overcome this barrier. These include processes where a first block comprising the MAM is prepared by atom transfer radical polymerization (ATRP), the ATRP initiator functionality (a bromine end group) is converted to a xanthate end group and the resultant macro-RAFT agent is then used in the synthesis of a polyMAM-*block*-polyLAM (*9, 10*); e.g., PSt-*block*-PNVP (*10*). Compounds that combine RAFT agent and ATRP initiator functionality in one molecule (*11–13*) have also been used to synthesize polyMAM-*block*-polyLAM; e.g., PSt-*block*-PVAc (*12*). There are reports of direct synthesis of polyMAM-*block*-polyLAM using other mechanisms (*14, 15*); e.g., stibine-mediated polymerization was used to synthesize PSt-*block*-PNVP (*14*) and a combination of cobalt mediated polymerization and ATRP was used to synthesize PSt-*block*-PVAc (*15*). Finally, copper-catalyzed 1,3-dipolar addition was used to link RAFT-synthesized polyLAM and polyMAM blocks including PSt-*block*-PVAc (*16, 17*).

In two recent communications (*18, 19*) we reported on a new class of stimuli-responsive RAFT agents, the *N*-4-pyridinyl-*N*-methyldithiocarbamates (**1-4,** Scheme 1) that can be "switched" to offer good control over polymerization of both MAMs and LAMs and thus a more direct and convenient route to polyMAM-*block*-polyLAM with narrowed molecular weight distributions. This paper more fully reports those results and presents additional data.

Results and Discussion

The *N*-4-pyridinyl-*N*-methyldithiocarbamates (**1-4**) provide effective control over the polymerization of VAc and other LAMs (NVP, NVC – see Table 1) but give at best poor control over the polymerization of MAMs. In this, their activity is similar to *N*-phenyl-*N*-methyldithiocarbamates (*20*). However, in the presence of a strong acid, the protonated form of the *N*-(4-pyridinyl)-*N*-methyldithiocarbamates (**1-H$^+$-4-H$^+$**) give excellent control over the polymerization of MAMs (MA, BA, MMA, St – see Table 1). The protonated RAFT agents strongly retarded or inhibited the polymerization of LAMs. The activity of **1-H$^+$-4-H$^+$** resembles that of aromatic dithiocarbamates, such as the *N*-pyrrolecarbodithioates (*21*).

The *N*-4-pyridinyl-*N*-methyldithiocarbamates can be synthesized using standard procedures (*4, 6*). Those with primary (**1**) or secondary "R" substituents (**2**) were synthesized in high yield by alkylation of the carbodithioate anion. Tertiary derivatives **3** and **4** were prepared by reaction of the dithiuram disulfide (**5**) with the appropriate initiator [azobis(isobutyronitrile) - AIBN or 2,2′-azobis(4-methoxy-2,4-dimethylvaleronitrile – V70 respectively] as shown in Scheme 2 (*22*).

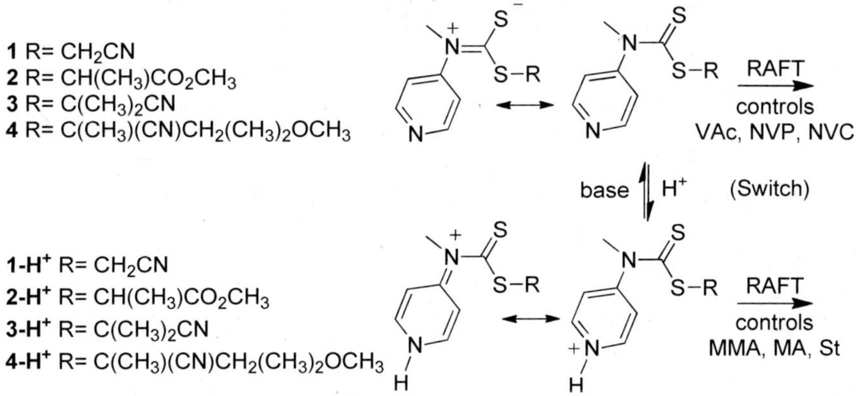

Scheme 1. Switchable RAFT agents.

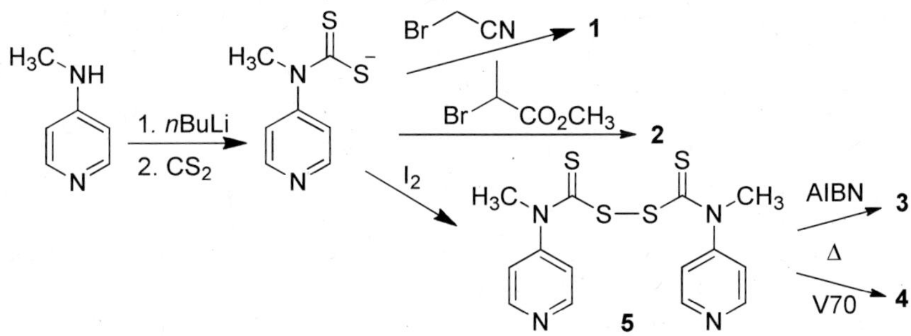

Scheme 2. Synthesis of N-4 pyridinyl-N-methyldithiocarbamates.

The efficiency of the protonated dithiocarbamate in controlling the polymerization of MAMs suggests the importance of the canonical form in which charge is localized on the dithiocarbamate nitrogen (Scheme 1). Molecular orbital calculations (AM1) provide support for this hypothesis (Table 2). They predict a decrease in C=S bond length, the energy LUMO and the partial charge on the thiocarbonyl sulfur for the protonated RAFT agent with respect to the non-protonated form.

The acid used to form the protonated RAFT agent should be a strong acid such as 4-toluenesulfonic acid (TsOH) or trifluoromethanesulfonic acid, such that the pyridine is fully ptotonated, and the acid should be added in a stoichiometric amount. The use of excess acid may cause hydrolysis of the RAFT agent while the use of less than stoichiometric acid or use of a weaker acid, such as acetic acid, provides poorer control which is manifest as a slighly broader molecular weight distribution. Non-protic Lewis acids such as aluminum triflate also give effective switching. The facile reversibility of protonation and the stability of the RAFT agent **2** to the switching process has been demonstrated by ^{1}H NMR (Figure 1) (*18*).

Table 1. Details of Polymer Syntheses (for further information see Experimental)

RAFT Agent /acid	*10^{-2} M*	*M1[a]*	*M*	*Temp °C*	*Solvent[a]*	*Initiator[a]*	*10^{-2} M*	*$\bar{M}_n$ 10^{3b}*	*Đ[c]*	*Time h*	*Conv %*
2	2.47	**NVC**	2.09	60	dioxane	AIBN	1.62	15.8	1.09	20	80
2-H⁺	2.47	**NVC**	2.09	60	dioxane	AIBN	1.62	oligomeric product			
2	1.25	**NVP**	4.68	60	CH_3CN	AIBN	0.30	29.4	1.19	16	83
1	2.84	**VAc**	7.21	75	EtAc	ACHN	0.16	9.6	1.23	40	82
2-H⁺ /TsOH	1.41	**MA**	4.44	70	CH_3CN	AIBN	0.04	31.1	1.08[d]	7	87
1	0.80[f]	**BA**	2.79	70	CH_3CN	AIBN	0.04	51.4	1.88	6	85
1-Al(OTf)₃	0.80	**BA**	2.79	70	CH_3CN	AIBN	0.04	47.0	1.14	6	85
2	1.35[f]	**BA**	2.79	70	CH_3CN	AIBN	0.04	21.3	1.57	6	84
2-H⁺ /TsOH	1.35	**BA**	2.79	70	CH_3CN	AIBN	0.04	24.1	1.12	6	91
3-H⁺ /TfOH	1.99	**MMA**	6.55	60	CH_3CN	AIBN	0.61	33.0	1.25[e]	16	98
4-H⁺ /TfOH	1.99	**MMA**	6.55	60	CH_3CN	AIBN	0.61	31.4	1.10	16	65

[a] Abbreviations TsOH = 4-toluenesulfonic acid, TfOH = trifluoromethanesulfonic acid, $Al(OTf)_3$ = aluminum trifluoromethanesulfonate, **MA** = methyl acrylate, **MMA** = methyl methacrylate, **St** = styrene, **VAc** = vinyl acetate, **NVC** = *N*-vinylcarbazole, **NVP** = *N*-vinylpyrrolidone, EtAc = ethyl acetate, AIBN=azobisisobutyronitrile, ACHN = azobis(1-cyclohexanenitrile). [b] Number average molecular weight in polystyrene equivalents. [c] Ratio of weight to number average molecular weight. [d] Refer Figure 2b. [e] Refer Figure 2a. [f] In the absence of acid there is control over molecular weight at high monomer conversion but molecular weight distributions are broader.

Table 2. HOMO and LUMO Energies, Partial Charges on Sulfur and C=S Bond Lengths for Thiocarbonylthio Compounds ZC(=S)S-CH_3[a]

Z	*Energy (eV)*		*atomic charge*		*bond length (Å)*	*ref*
	HOMO	*LUMO*	*=S*	*-S*	*C=S*	
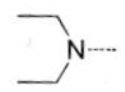	-0.842	-8.786	-0.222	0.168	1.59	(*8*)
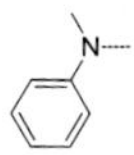	-0.571	-8.580	-0.208	0.204	1.59	this work
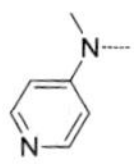	-0.713	-8.592	-0.186	0.209	1.58	this work
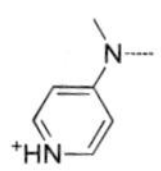	-0.919 [b]	-8.836 [b]	0.096	0.275	1.56	this work
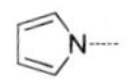	-1.395	-8.895	0.035	0.292	1.57	(*8*)

[a] Results of semi-empirical molecular orbital calculations (AM1). [b] Values of the HOMO and LUMO energies depend strongly on the presence, type and position of a counterion.

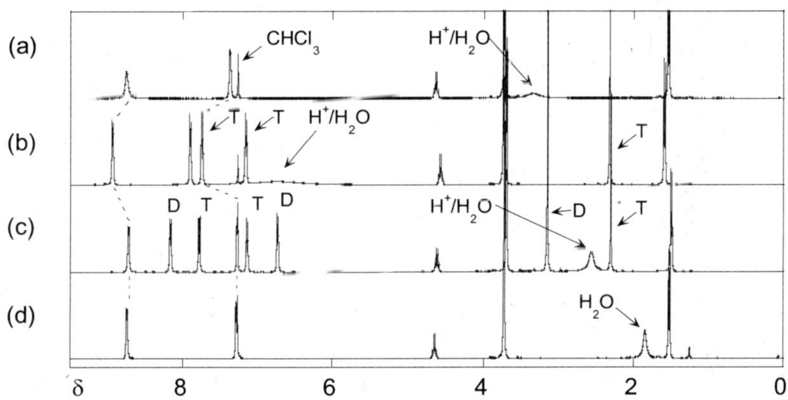

*Figure 1. 400 MHz ^{1}H NMR spectra ($CDCl_3$) of (a) RAFT agent **2** (b) with one equivalent of 4-toluenesulfonic acid (T), (c) with one equivalent of 4-dimethylaminopyridine (D) and (d) after aqueous extraction to remove the 2,6-dimethylpyridinium 4-toluenesulfonate.*

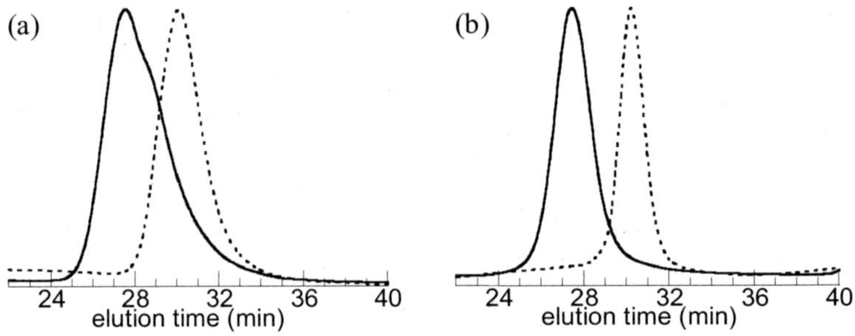

Figure 2. GPC chromatograms of (a) poly(methyl methacrylate) (M_n 33000, Đ 1.25) prepared with ***3-H*⁺** *(------) and poly(methyl methacrylate)-block-poly(N-vinyl acetate) (M_n 55900, Đ 1.39) (——), and (b) poly(methyl acrylate) (M_n 31100, Đ 1.08) prepared with* ***2-H*⁺** *(------) and poly(methyl acrylate)-block-poly(N-vinylcarbazole) (M_n 48000, Đ 1.33) (——). (Reproduced from ref. (18) © American Chemical Society).*

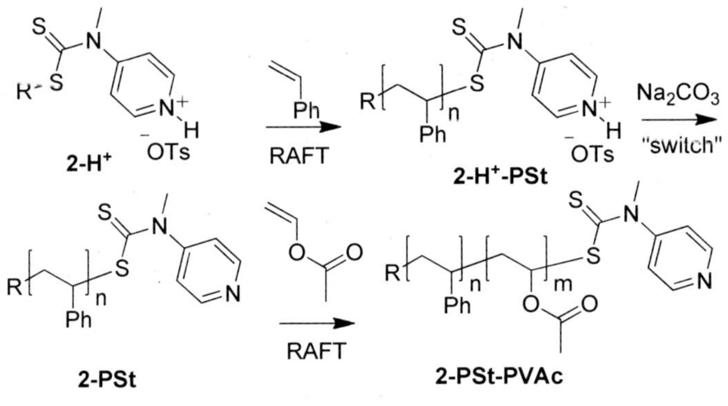

Scheme 3. Synthesis of polystyrene-block-poly(vinyl acetate) ($R=CH(CH_3)CO_2CH_3$).

The use of switchable RAFT agents in block copolymer synthesis was demonstrated with the use of 4-pyridinyl-*N*-methyldithiocarbamate derivatives to prepare PMMA-*block*-PVAc (*18*), PMA-*block*-PNVC (*18*) (Figure 2) and PS-*block*-PVAc (Table 3) (*19*). In applying the method to polyMAM-*block*-polyLAM, it was found that the block comprising MAMs needed to be synthesized first. This sequence is necessary because, polyLAM• are poor radical leaving groups with respect to polyMAM• and, consequently, polyLAM macro-RAFT agents have very low transfer constants in MAM polymerization.

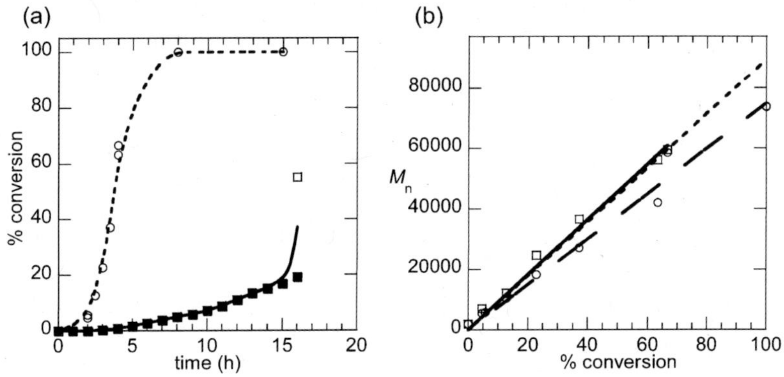

Figure 3. (a) Evolution of total monomer conversion vs time for bulk polymerization of vinyl with polystyrene macro-RAFT agent (0.0108 M) at 70 °C and 0.0057 M (------, ○) or 0.0011 M azobis(isobutyronitrile) (AIBN) (——, □, ■). Data points (■) from in situ ^{1}H NMR. (b) Evolution of number average molecular weight (M_n) with conversion for bulk polymerization of vinyl acetate with polystyrene macro-RAFT agent (0.0108 M) and AIBN (0.0057 M) at 70 °C; GPC M_n in polystyrene equivalents (— — —, ○), M_n from NMR analysis (——, □); calculated M_n (-------); lines are lines of best fit.

Attempts to synthesize PSt-*block*-PVAc or other PSt-*block*-polyLAM starting from a PSt macro-RAFT agent prepared with a 4-pyridinyl-*N*-methyldithiocarbamate using the RAFT polymerization conditions (initiator:RAFT agent ~ 1:10) that had proved successful for other polyMAM-*block*-polyLAM or PS-*block*-polyMAM gave only a very low yield of polymer after an extended period (>4 h; Figure 3a). This was attributed to the very low rate of PSt• addition to VAc, possibly compounded by the presence of trace amounts of St monomer in the PSt macroRAFT agent. St is known to inhibit the polymerization of VAc which is a consequence of adverse reactivity ratios (r_S=22.3, r_{VAc}=0.02) (*23*). The rate constants for benzyl radical (a simple model for PSt•) addition to St and VAc are 1100 and 15 M s^{-1} respectively at 25 °C (*24*).

Several solutions have been found to this dilemma. The first was to use a larger than usual ratio of initiator to RAFT agent (e.g., 1:2). The premise is that once the initial macro-RAFT agent has been converted to a macro-RAFT agent with VAc as the terminal unit polymerization should proceed normally. In the first step (Scheme 3), the protonated RAFT agent **2-H⁺** was used to control the polymerization of St to form **2-H⁺-PSt**. This macro-RAFT agent was then neutralized, forming **2-PSt**, and used in VAc polymerization to provide the desired **2-PSt-PVAc**. A relatively short period (<2 h) of retarded polymerization was observed, while the **2-PSt** was converted to **2-PSt-PVAc,** after which rapid polymerization was observed (Figure 3a).

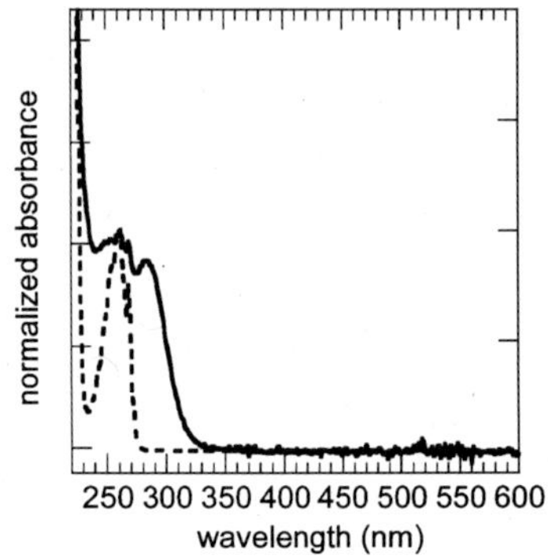

Figure 4. Ultraviolet spectra ($CHCl_3$) of (a) polystyrene-block-poly(vinyl acetate) ***1-PSt-VAc*** *prepared by RAFT polymerization extracted from GPC chromatogram at retention time corresponding to peak maximum (26 min) (——) and (b) polystyrene (Polymer Laboratories) (- - -).*

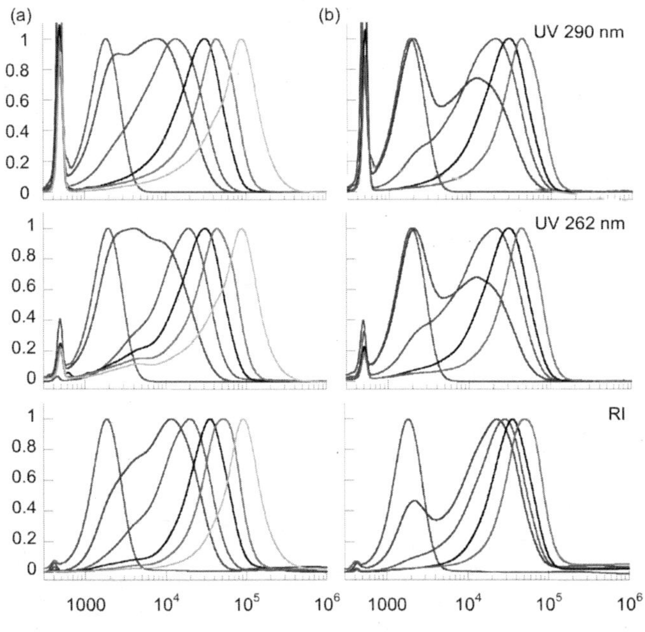

Figure 5. Evolution of normalized molecular weight distributions with conversion for (a) polystyrene-block-poly(vinyl acetate) obtained at 0, 2, 2.5, 3, 3.5 and 4 h (left to right) and (b) polystyrene-block-poly(methyl acrylate-grad-vinyl acetate)-block-poly(vinyl acetate) obtained with: (RI) with refractive index detector, (UV 262) with UV detector at 262 nm and (UV 290) at 290 nm. (Reproduced from ref. (19) © American Chemical Society).

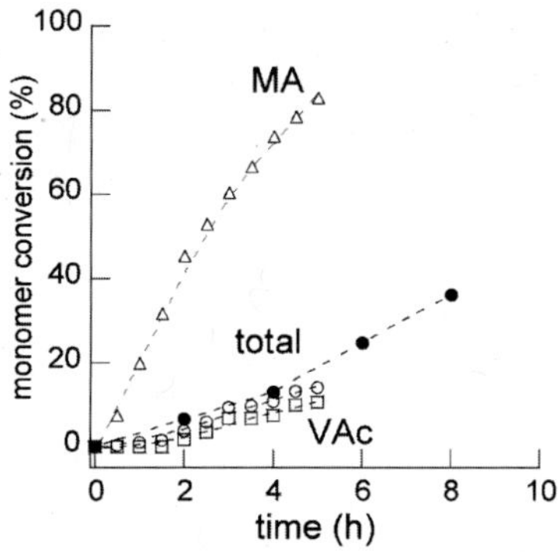

Figure 6. Evolution of monomer conversion vs time for bulk copolymerization of methyl acrylate (MA) and vinyl acetate (VAc) (1:20) with polystyrene macroRAFT agent at 70 °C. VAc conversion determined by NMR (- -, □). MA conversion determined by NMR (- -, Δ). Total conversion determined by NMR (- -, ○). Total conversion determined gravimetrically (- -, ●)

The evolution of the molecular weight distribution with time in this experiment was followed with a GPC equipped with both a refractive index (RI) and a diode array UV detector. The traces extracted at 262 nm where the polystyrene aromatic ring chromophore has an absorption maximum (Figure 4), and at 290 nm, where the thiocarbonylthio chromophore has an absorption maximum and there is minimal absorption by the polystyrene, are shown in Figure 5a. There was close correspondence between the found and calculated molecular weights (Figure 3b). However, the formation of some dead PSt formed by combination of PSt• radicals from **2-PSt** is clearly evident in the UV262 distribution but not in the UV290 distribution and is also less evident in the RI distribution which is dominated by PVAc for high conversion samples after the first time point. The presence of dead polystyrene nonetheless accounts for the high molecular weight dispersity (*Đ*) of **2-PSt-PVAc** formed in this experiment (Table 3). It is extremely important, when applying this strategy to prepare PSt-*block*-polyLAM, to ensure that the PSt macroRAFT agent is completely free of residual St monomer otherwise the high reactivity of St will inevitably result in formation of new macroRAFT agent with a terminal St unit and an extended period of slow polymerization.

Other solutions follow from the observation that it was possible to synthesize both low *Đ* PSt-*block*-PMA and PMA-*block*-PVAc (*18*). Thus, PSt-*block*-PMA-*block*-PVAc was successfully prepared in three steps with isolation of PSt-*block*-PMA (**2-PSt-MA**, Table 3). In principle the MA "block" need only be a single unit in length.

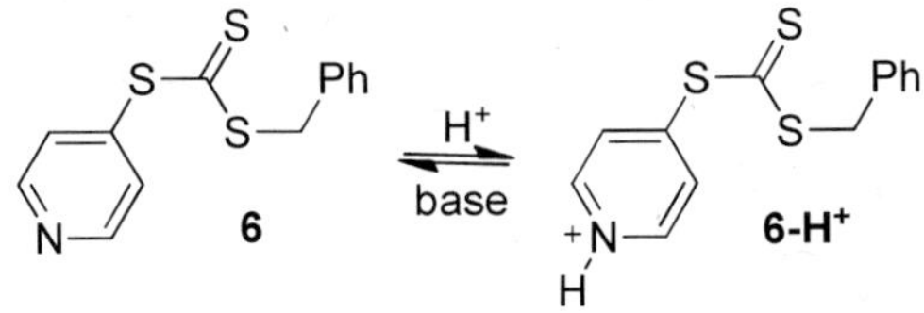

Scheme 4. Benzyl pyridine-4-yl trithiocarbonate (benzyl pyridin-4-yl carbonotrithioate).

A more convenient process is to copolymerize a mixture of MA and VAc in the presence of the PSt macroRAFT agent (**2-PSt**) in its deprotonated form (Table 3). Since the deprotonated **2-PSt** has only a low transfer constant in MA polymerization (*18*) it is consumed slowly. However, the higher reactivity of MA towards both PSt• and PVAc• dictates that PSt-*block*-P(MA-*grad*-VAc)-*block*-PVAc (*25*, *26*) should be formed and that polymerization proceeds with no discernable inhibition period and without significant formation of dead PSt (as shown by close correspondence of UV262 and UV 290 traces in Figure 5b). In the copolymerization experiment the MA:VAc ratio was 1:20 and the NMR copolymer analysis is consistent with the MA, and the initial macroRAFT agent, being largely consumed before a total monomer conversion of 25% (Figure 6).

We have also examined the similar trithiocarbonate as a RAFT agent (*27*). Both the protonated (**6,** Scheme 4) and unprotonated form (**6-H⁺**) are only effective in controlling the polymerization of MAMs. Values of *Đ* for < 4 h reaction time in RAFT polymerizations of St at 70 °C with **6-H⁺** (trifluoroacetate) were significantly lower than with **6** though the polymerizations with **6-H⁺** were slightly retarded with respect to the similar polymerization with the **6**. Consistent with this observation, the transfer constant (C_{tr}^{app}) in St polymerization, estimated on the basis of the residual RAFT agent, for **6-H⁺** RAFT agent appeared ~25% higher (37 for **6-H⁺** *vs* 30 for **6**) (*27*).

Conclusions

We have discovered a new class of "switchable" RAFT agents. The *N*-(4-pyridinyl)-*N*-methyldithiocarbamates provide excellent control over polymerization of LAMs and after addition of one equivalent of a protic or Lewis acid they become effective in controlling polymerization of MAMs. The "switchable" RAFT agents allow the synthesis of polyMAM-*block*-polyLAM also with narrow molecular weight distributions and was successfully demonstrated with the synthsis of PMMA-*block*-PVAc and PMA-*block*-PNVC. However, because of the very low reactivity of PSt• in VAc polymerization, the synthesis of block copolymers of PSt and PVAc is best accomplished by forming a PSt-*block*-PMA as an intermediate.

Table 3. Details of Polymer Syntheses (for further information see Experimental)

E[a]	*RAFT Agent*[a]	*10-2 M*[b]	*Mon*[c]	*M*[b]	*Temp °C*	*Solvent*	*Init*[c]	*10-2 M*[b]	M_n */10*[3df]	*Đ*[ef]	*Time h*	*Conv %*
1	**2-H**[+g]	7.0	**St**	8.75	90	bulk	ACHN	1.01	1.8(1.9)	1.22(1.11)	2.0	18
2	**2-PSt**	1.08	**VAc**	10.86	70	bulk	AIBN	0.57	5.4(7.9)	2.16(1.81)	2.0	4.6
3	**2-PSt**	1.08	**VAc**	10.86	70	bulk	AIBN	0.57	10.4(13.3)	1.91(1.62)	2.5	13
4	**2-PSt**	1.08	**VAc**	10.86	70	bulk	AIBN	0.57	18.2(27.0)	2.00(1.42)	3.0	23
5	**2-PSt**	1.08	**VAc**	10.86	70	bulk	AIBN	0.57	27.7(39.9)	1.90(1.40)	3.5	37
6	**2-PSt**	1.08	**VAc**	10.86	70	bulk	AIBN	0.57	50.2(59.5)	1.83(1.61)	4.0	67
7	**2-PSt**	1.08	**VAc**	10.86	70	bulk	AIBN	0.11	52.4	1.37	16.0	55
8	**2-H**[+g]	7.02	**St**	8.75	90	bulk	ACHN	1.1	10.5	1.06	12.0	82
9	**2-PSt-H**[+g]	1.06	**MA**	5.52	70	benzene	AIBN	1.0	21.5	1.15	2.0	25
10	**2-PSt-PMA**	0.05	**VAc**	10.86	70	bulk	AIBN	0.50	52.0	1.48	15.0	75
11	**2-PSt**	1.08	**VAc/MA**	10.3/0.55	70	bulk	AIBN	0.11	6.6	2.90	2.0	6.6
12	**2-PSt**	1.08	**VAc/MA**	10.3/0.55	70	bulk	AIBN	0.11	14.1	2.04	4.0	13
13	**2-PSt**	1.08	**VAc/MA**	10.3/0.55	70	bulk	AIBN	0.11	25.1	1.58	6.0	25
14	**2-PSt**	1.08	**VAc/MA**	10.3/0.55	70	bulk	AIBN	0.11	42.7	1.31	8.0	36

[a] The macro-RAFT agent formed in Expt. 1 was neutralized and used in Expts. 2-7 and 11-14. That formed in Expt. 8 was used in Expt. 9. That formed in Expt. 9 was neutralized and used in Expt. 10. [b] Concentrations at 22 °C. [c] Monomer: **MA**=methyl acrylate, **St**=styrene, **VAc**=vinyl acetate. Initiator: AIBN=azobisisobutyronitrile, ACHN=azobis(1-cyclohexanenitrile). [d] Number average molecular weight in polystyrene equivalents from conventional analysis of RI trace. [e] Ratio of weight average to number average molecular weight. [f] M_n and *Đ* for **2-PSt-PVAc** from analysis of UV290 trace is given in parentheses (excludes contribution from dead polymer). [g] Acid used was 4-toluenesulfonic acid.

Experimental

General

Solvents were of AR grade and were distilled before use. Monomers, MA, BA, MMA, S, and VAc were obtained from Aldrich and were filtered through neutral alumina (70-230 mesh), fractionally distilled under reduced pressure, and flash distilled under reduced pressure immediately before use. NVP (Aldrich) was purified by vacuum distillation. NVC (Pfaltz & Bauer) was used as received. Initiators, AIBN (DuPont VAZO-64®), azobis(cyclohexanenitrile) (ACHN, DuPont VAZO-88®), were purified by crystallization from chloroform/methanol. V70 (Wako) was used as received. Gel permeation chromatography (GPC) was performed with a Waters Associates liquid chromatograph equipped with differential refractometer and 3×mixed C and 1 mixed E PLgel column (each 7.5 mm×300 mm) from Polymer Laboratories. Tetrahydrofuran (flow rate of 1.0 mL/min) was used as eluent at 22±2 °C. The columns were calibrated with narrow dispersity PSt standards (Polymer Laboratories). A third order polynomial was used to fit the $\log_{10}M$ *vs* time calibration curve, which appeared approximately linear across the molecular weight range 2×10^2 - 2×10^6 g mol^{-1}. The molecular weights in this paper are reported as PSt equivalents. Samples for GPC analysis were isolated by evaporation of solvent and unreacted monomer. No precipitation or fractionation was performed prior to GPC analysis. GPC of PNVP and PNVC was performed on a system comprising a Waters 590 HPLC pump and a Waters 410 refractive index detector equipped with 3×Waters Styragel columns (HT2, HT3, HT4 each 300 mm×7.8 mm providing an effective molecular weight range of 100-600000). The eluent was *N,N*-dimethylformamide (containing 0.045% w/v LiBr) at 80 °C (flow rate: 1 mL min^{-1}). Quoted monomer conversions were determined gravimetrically unless stated otherwise. The NMR spectra were recorded on a Bruker AC200 (200 MHz for ^{1}H NMR) or Bruker Av400 spectrometer (400 MHz for ^{1}H NMR) where indicated. Chemical shifts are quoted relative to external tetramethylsilane. High resolution electron impact (HREI) mass spectra (MS) were obtained with a ThermoQuest MAT95XP mass spectrometer employing electron impact (EI) at 70eV and with perfluorokerosene as a reference.

RAFT Agent Synthesis

*Cyanomethyl Methyl(pyridin-4-yl)carbamodithioate (**1**)*

n-Butyl lithium (2.5 M in hexane, 44 mL, 0.11 mol) was added dropwise to a cold (-10 °C) solution of 4-(methylamino)pyridine (9 g, 0.0841 mol) in anhydrous THF (200 mL) at a rate such that the temperature remained below -7 °C. The resulting pale yellow suspension was then allowed to stir for 1 h at -10 °C when carbon disulfide (10.2 mL, 12.936 g, 0.168 mol) was added over 1 h at 0 °C and the mixture was left to stir overnight at room temperature. The resultant mixture was cooled to 0 °C and bromoacetonitrile (8.8 mL, 15.14 g, 0.126 mol) was added dropwise. The mixture was then stirred at room temperature for a further 2 h. The reaction mixture was extracted with diethyl ether (400 mL), washed with saturated

$NaHCO_3$ and brine. The organic layers were dried (Na_2SO_4) and reduced to a dark brown oil under vacuum. The crude was dissolved in ethyl acetate and purified by column chromatography (silica gel 60, 70-230 mesh, 30 vol% ethyl acetate in *n*-hexane as eluent) to give **1** as off-white cubes (11.05 g, 59% yield). ^1H NMR ($CDCl_3$) δ 3.78 (s, 3H, N-C**H**$_3$); 4.05 (s, 2H, SC**H**$_2$CN); 7.25 (m, 2H, *m*-Ar**H**); 8.75 (m, 2H, *o*-Ar**H**). MS (HREI) m/z 223.0230 (M^+) ($C_9H_9N_2S_2$ requires 223.0238).

Methyl 2-((Methyl(pyridin-4-yl)carbamothioyl)thio)propanoate (2)

n-Butyl lithium (1.6M in hexane) (6 M in hexane, 6.25 mL, 10 mmol) was added over 15 minutes to a cold (–10 °C) solution of 4-(methylamino)pyridine (1.08 g; 10 mmol) in dry THF (60 mL) over 15 minutes under nitrogen. The resultant pale yellow mixture was then allowed to stir at -10 °C for 1 h when it was warmed to 0 °C and carbon disulfide (0.9 mL) was added dropwise. The yellow suspension was then allowed to stir at room temperature overnight. Methyl 2-bromopropionate (1.23 mL, 11 mmol) was then added dropwise and stirring was continued for 2 hours. The solvent THF was removed under vacuum, the residue suspended in ethyl acetate and filtered. After removal of the ethyl acetate under vacuum the crude product was purified by flash chromatography with ethyl acetate as eluent to afford methyl 2-((methyl(pyridin-4-yl)carbamothioyl)thio)propanoate (**2**) as an off-white solid (2.3 g, 85.4% yield). ^1H NMR ($CDCl_3$) δ 1.55 (d, 3H, CHC**H**$_3$); 3.70 (s, 6H, COOC**H**$_3$ and N-C**H**$_3$); 4.65 (q, 1H, C**H**CH$_3$); 7.35 (d, 2H, *m*-Ar**H**); 8.75 (br s, 2H, *o*-Ar**H**). MS (HREI) m/z 270.0486 (M^+) ($C_{11}H_{14}N_2O_2S_2$ requires 270.0497).

Dithiuram Disulfide (5)

n-Butyl lithium (1.6 M in hexane, 6.5 mL) was added to a cold (–10 °C) solution of 4-(methylamino)pyridine (1.07g; 9.89 mmol) in dry THF (60 mL) over 15 minutes under nitrogen. The resultant pale yellow mixture was allowed to stir at -10 °C for 1 h and then warmed to 0 °C before the dropwise addition of carbon disulfide (0.9 mL) and the yellow suspension was allowed to stir at room temperature for further one hour. The resultant mixture was cooled to 0 °C and oxidized by the addition of iodine (1.23 g iodine in 25 mL of 10% aqueous KI) to yield dithiuram disulfide **5** (1.36 g, 75% yield). ^1H NMR ($CDCl_3$) δ 3.80 (s, 6H, 2×N-C**H**$_3$); 7.45 (d, 4H, *m*-Ar**H** x2); 8.75 (d, 4H, *o*-Ar**H** x2).

2-Cyanopropan-2-yl Methyl(pyridin-4-yl)carbamodithioate (3)

A solution of AIBN (0.206 g, 1.26 mmol) and dithiuram disulfide (**5**) (0.23 g, 0.628 mmol) in ethyl acetate (10 mL) was heated at reflux for 16 h. After removal of the volatiles under vacuum, the crude product was subjected to column chromatography (Kieselgel-60, 70-230 mesh), with *n*-hexane:ethyl acetate (1:1) as eluant to afford 2-cyanopropan-2-yl methyl(pyridin-4-yl)carbamodithioate (**3**) as an off-white liquid (0.31 g, 98.4% yield), which solidified during storage at -15 °C. ^{1}H NMR ($CDCl_3$) δ 1.82 (s, 6H, C(CN)(**CH_3**)$_2$); 3.65 (s, 3H, NC**H_3**); 7.20 (d, 2H, *m*-Ar**H**); 8.75 (br s, 2H, *o*-Ar**H**) . MS (HREI) m/z 251.0545 (M^+) ($C_{11}H_{13}N_3S_2$ requires 251.0551).

2-Cyano-4-methoxy-4-methylpentan-2-yl Methyl(pyridin-4-yl)carbamodithioate (4)

A solution of V70 (0.74 g, 2.4 mmol) and dithiuram disulfide (**5**) (0.73 g, 2.0 mmol) in ethyl acetate (20 mL) was heated at reflux for 16 h. After removal of the volatiles under vacuum, the crude product was subjected to column chromatography (Kieselgel-60, 70-230 mesh), with *n*-hexane:ethyl acetate (1:49) as eluant to afford 2-cyano-4-methoxy-4-methylpentan-2-yl methyl(pyridin-4-yl)carbamodithioate (**4**) as a pale yellow liquid (0.82 g, 63.5% yield), which solidified during storage at -15 °C. ^{1}H NMR ($CDCl_3$) δ 1.20 (s, 3H, **CH_3**); 1.40 (s, 3H, **CH_3**); 1.85 (s, 3H, C(CN)**CH_3**); 1.95 (d, 1H, -**CH_2**-); 2.80 (d, 1H, -**CH_2**-); 3.10 (s, 3H, O**CH_3**); 3.65 (s, 3H, N**CH_3**); 7.30 (d, 2H, *m*-Ar**H**); 8.70 (br s, 2H, *o*-Ar**H**).

Polymerizations

Preparation of Poly(methyl methacrylate) Using 2-Cyanopropan-2-yl methyl(pyridin-4-yl)carbamodithioate (3) and Trifluoromethanesulfonic Acid

A stock solution (I) of trifluoromethanesulfonic acid (100 μL, 170 mg) in acetonitrile (5.0 mL) was prepared. A stock solution (II) containing MMA (7.0 mL), AIBN (10 mg), 2-cyanopropan-2-yl methyl(pyridin-4-yl)carbamodithioate (**3**) (50.02 mg, 0.0199 M), acetonitrile (2.0 mL) and stock solution (I) (1.0 mL) was prepared. Aliquots (2.0 mL) of this stock solution (II) were transferred to ampoules, degassed through three repeated freeze-evacuate-thaw cycles and sealed. The ampoules were heated at 60 °C for the times indicated in the Table 4.

Preparation of Poly(methyl acrylate) Using Methyl 2-((Methyl(pyridin-4-yl)carbamothioyl)thio)propanoate (2) with 4-Toluenesulfonic Acid

A stock solution (I) of AIBN (8.5 mg) in acetonitrile (25 mL) was prepared. A stock solution (II) of **2** (47.6 mg, 0.0353 M) and TsOH (38.0 mg, 0.04M) in acetonitrile (5.0 mL) was prepared. Aliquots of stock solution (I) (1.0 mL), stock solution (II) (2.0 mL) and MA (2.0 mL) were transferred to ampoules which were degassed by three repeated freeze-evacuate-thaw cycles and sealed. The ampoules were heated at 70 °C for the times indicated in the Table 5.

Table 4. Molecular weight and conversion data for poly(methyl methacrylate) prepared with 2-cyanopropan-2-yl methyl(pyridin-4-yl)carbamodithioate (3) and trifluoromethanesulfonic acid at 60°C

Expt.	*time/h*	M_n[a]	*Đ*[b]	*% Conv.*
1	3	15500	1.56	27.6
2	6	19200	1.58	51.1
3	16	33100	1.25	98.0

[a] Number average molecular weight in polystyrene equivalents. [b] Ratio of weight average to number average molecular weight.

Table 5. Molecular weight and conversion data for poly(methyl acrylate) prepared with methyl 2-((methyl(pyridin-4-yl)carbamothioyl)thio)propanoate (2) and 4-toluenesulfonic acid at 70 °C

Expt.	*time/hr*	M_n[a]	*Đ*[b]	*% Conv.*
1	2	20900	1.09	58.9
2	7	31100	1.08	87.3

[a] Number average molecular weight in polystyrene equivalents. [b] Ratio of weight to number average molecular weight.

*Preparation of Poly(n-butyl acrylate) Using Methyl 2-((Methyl(pyridin-4-yl)carbamothioyl)thio)propanoate (**2**) with and without 4-Toluenesulfonic Acid*

A stock solution (I) of AIBN (8.5 mg) in acetonitrile (25 mL) was prepared. A stock solution (II) of methyl 2-((methyl(pyridin-4-yl)carbamothioyl)thio)propanoate (**2**) (47.6 mg, 0.0353 M) and TsOH (38.0 mg, 0.04 M) in acetonitrile (5.0 mL) was prepared. A stock solution (III) of methyl 2-(methyl(pyridin-4-yl)carbamothioylthio)propanoate (**2**) (47.6 mg, 0.0353 M) in acetonitrile (5.0 mL) was prepared. Aliquots of stock solution I (1.0 mL), stock solution II (2.0 mL) and *n*-butyl acrylate (2.0 mL) were transferred to ampoules (for Table 6, Expts. 1 and 2). Aliquots of stock solution I (1.0 mL), stock solution III (2.0 mL) and *n*-butyl acrylate (2.0 mL) were transferred to ampoules (for Table 6, Expts. 3 and 4). The contents of the ampoules were degassed by three repeated freeze-evacuate-thaw cycles, sealed and heated at 70 °C for the times indicated in the Table 6.

Table 6. Molecular weight and conversion data for poly(*n*-butyl acrylate) prepared methyl 2-((methyl(pyridin-4-yl)carbamothioyl)thio)propanoate (2) with 4-toluenesulfonic acid at 70 °C

Expt.	*time/hr*	M_n[a]	$Đ$[b]	*% Conv.*
1	2	14300	1.19	51.7
2	6	24200	1.12	91.9
3[c]	2	19800	1.73	53.7
4[c]	6	21300	1.57	84.3

[a] Number average molecular weight in polystyrene equivalents. [b] Ratio of weight average to number average molecular weight. [c] Control experiment without 4-toluenesulfonic acid

Table 7. Molecular weight and conversion data for poly(*n*-butyl acrylate) prepared with cyanomethyl methyl(pyridin-4-yl)carbamodithioate (1) with and without Lewis acid aluminium trifluoromethanesulfonate at 70 °C

Expt.	*time/hr*	M_n[a]	$Đ$[b]	*% Conv.*
2A	2	31000	1.16	59.2
6A	6	47040	1.14	84.5
2B	2	51500	1.74	65.1
6B	6	51400	1.88	84.7

[a] Number average molecular weight in polystyrene equivalents. [b] Ratio of weight average to number average molecular weight.

*Preparation of Poly(n-butyl acrylate) Using Cyanomethyl Methyl(pyridin-4-yl)carbamodithioate (**1**) with and without Aluminum Trifluoromethanesulfonate*

Stock solutions (I) of AIBN (8.5 mg) in acetonitrile (25 mL), (II) of cyanomethyl methyl(pyridin-4-yl)carbamodithioate (**1**) (22.3 mg) and aluminium trifluoromethanesulfonate (47.4 mg) in acetonitrile (5.0 mL) and (III) of cyanomethyl methyl(pyridin-4-yl)carbamodithioate (**1**) (22.3 mg) in acetonitrile (5.0 mL) were prepared. Aliquots of stock solution I (1.0 mL), stock solution II (2.0 mL) and *n*-butyl acrylate (2.0 mL) were transferred to two ampoules labelled 2A and 6A. Aliquots of stock solution I (1.0 mL), stock solution III (2.0 mL) and *n*-butyl acrylate (2.0 mL) were transferred to two ampoules labelled 2B and 6B. The contents in ampoules were degassed by three freeze-evacuate-thaw cycles, sealed and heated at 70 °C for the times indicated in the Table 7.

Table 8. Molecular weight and conversion data for poly (vinyl actetate) prepared with methyl 2-((methyl(pyridin-4-yl)carbamothioyl)thio)propanoate (2) at 75 °C

Expt.	*time/hr*	$M_n{}^a$	$Đ^b$	*% Conv.*
1	24	1000	1.39	10.1
2	40	9640	1.23	82.1

[a] Number average molecular weight in polystyrene equivalents. [b] Ratio of weight average to number average molecular weight.

*Preparation of Poly(vinyl acetate) Using S-Cyanomethyl N-Methyl, N-(Pyridin-4-yl)carbamodithioate (**1**)*

A stock solution of *S*-cyanomethyl *N*-methyl, *N*-(pyridin-4-yl)carbamodithioate (**1**) (253.4 mg), VAZO-88 (15.2 mg), VAc (15 mL) and ethyl acetate (5.0 mL) was prepared. Aliquots (5.0 mL) of this stock solution were transferred to ampoules which were degassed by three freeze-evacuate-thaw cycles and sealed. The ampoules were heated at 75 °C for the specified times. After the reaction, the un-reacted monomer was removed on rotary evaporator. The results of these experiments are presented in Table 8.

*Preparation of Poly(N-vinyl pyrrolidone) Using Methyl 2-((Methyl(pyridin-4-yl)carbamothioyl)thio)propanoate (**2**)*

A stock solution (I) comprising of AIBN (10.0 mg) in acetonitrile (5 mL) was prepared. *N*-vinyl pyrrolidone (2.0 mL), stock solution I (1.0 mL) and methyl 2-((methyl(pyridin-4-yl)carbamothioyl)thio)propanoate (**2**) (13.5 mg) in acetonitrile (1.0 mL) were transferred into an ampoule which was degassed by three freeze-evacuate-thaw cycles and sealed. The ampoule was heated at 60 °C for 16 h. After the reaction, removed the organic solvent, and the residue was added slowly into ethyl acetate (200 mL) to isolate low dispersity poly(*N*-vinyl pyrrolidone) (1.73 g, 82.9% conversion). GPC (DMF eluent): M_n 29,400; M_w/M_n 1.19

*Preparation of Poly(N-vinyl carbazole) Using Methyl 2-((Methyl(pyridin-4-yl)carbamothioyl)thio)propanoate (**2**)*

NVC (0.5 g), AIBN (2.0 mg) and methyl 2-((methyl(pyridin-4-yl)carbamothioyl)thio)propanoate (**2**) (5.0 mg) in 1,4-dioxane (0.75 mL) were transferred into a test-tube ampoule which was degassed by three freeze-evacuate-thaw cycles and sealed. The ampoule was heated at 60 °C for 20 h. After the reaction, the reaction mixture was added slowly into *n*-hexane (150 mL) to isolate low dispersity poly(*N*-vinyl carbazole) (0.398 g, 79.7% conversion). GPC (DMF eluent): M_n 15,800; M_w/M_n 1.09

Table 9. Molecular weight and conversion data for polystyrene prepared with methyl 2-((methyl(pyridin-4-yl)carbamothioyl)thio)propanoate (2) and 4-toluenesulfonic acid at 90°C

Expt.	*[S]₀* *M*	*[I]₀* *10^{-2} M*	*[2]₀* *10^{-2} M*	*[TsOH]* *10^{-2} M*	*time* *h*	M_n[a]	$Đ$[b]	*% Conv.*
1	8.7[c]	1.1	7.02	7.01	2	1800	1.33	19.5
2	8.7[c]	1.1	7.02	7.01	6	6000	1.08	47.2
3	8.7[c]	1.1	7.02	7.01	12	10500	1.06	81.6
5[d]	8.7[c]	1.1	7.02	7.01	2	1850	1.22	18.0
6	7.3[e]	0.9	5.84	5.83	3	2400	1.14	25.0

[a] Number average molecular weight in polystyrene equivalents. [b] Ratio of weight average to number average molecular weight. [c] Bulk styrene. [d] Duplicate of Expt 1. [e] Acetonitrile solution.

Preparation of Poly(methyl acrylate)-b-Poly(N-vinylcarbazole)

A stock solution (I) consisting of the PMA (1.67 g, M_n 31,100; M_w/M_n 1.08), *N,N*-dimethylaminopyridine (10.0 mg) in acetonitrile (10 mL) was prepared. A stock solution (II) of AIBN (10 mg) in acetonitrile (5 mL) was prepared. *NVC* (0.5 g), stock solution (I) (2.0 mL), stock solution (II) (1.0 mL) and acetonitrile (1.0 mL) were transferred into a test-tube ampoule which was degassed by three freeze-evacuate-thaw cycles and sealed. The ampoule was heated at 60 °C for 16 hours. The volatiles were removed *in vacuo* to give PMA-*b*-poly(*N*-vinyl carbazole) (almost complete conversion of *N*-vinyl carbazole based on ^{1}H-NMR) with GPC result M_n 48,000, M_w/M_n 1.33.

Preparation of Polystyrene Macro-RAFT Agent (2-PSt) with Methyl 2-((Methyl(pyridin-4-yl)carbamothioyl)thio)propanoate (2)

A solution comprising ACHN (12.2 mg), methyl 2-((methyl(pyridin-4-yl)carbamothioyl)thio)propanoate (**2**) (94.5 mg), TsOH (60.2 mg) and St (5 mL) was transferred to an ampoule which was degassed by three freeze-evacuate-thaw cycles, sealed, then heated at 90 °C as indicated (Table 9). To avoid unwanted cationic polymerization of St, it is important that TsOH is not added directly to St monomer. The ampoule was cooled, opened and the polymerization mixture evaporated to dryness *in vacuo*. The residue was taken up in chlorobenzene and evaporated under vacuum several times until no St signals were detectable by ^{1}H NMR. The residue (**2-PSt-H⁺**) was dissolved in dichloromethane and percolated through a carefully crushed and dried sodium carbonate bed. The color of the solution changed from yellow to colorless. Removal of the solvent gave the PSt macroRAFT (**2-PSt**) agent ready to be used.

Table 10. Molecular weight and conversion data for bulk vinyl acetate polymerization in presence of polystyrene macro-RAFT agent (2-PSt) (0.0108 M) with azobis(isobutyronitrile) (0.0057 M) at 70 °C

Expt.	M_n $/10^{3a}$	${}^{NMR}M_n$ $/10^{3b}$	$M_n(calc)$ $/10^{3c}$	$Đ^d$	*Time/h*	*Conv* %
1	5.4	7.1	5.9	2.16	2.0	4.6
2	10.4	12.2	13.0	1.91	2.5	13
3	18.2	24.6	21.8	2.00	3.0	23
4	27.7	33.6	34.1	1.90	3.5	37
5	42.0	56.0	56.9	1.83	4.0	63
6	58.4	59.4	59.7	1.73	4.0	67
7	73.6	-	87.8	1.95	8.0	100
8	73.7	-	87.8	2.13	15.0	100

[a] Number average molecular weight in polystyrene equivalents. [b] Molecular weight determined by NMR analysis. [c] Calculated molecular weight = 1850 + Conv×86.9×($[VAc]_0$/$[\mathbf{2\text{-}PSt}]_0$). [d] Ratio of weight average to number average molecular weight.

Preparation of Polystyrene-block-Poly(vinyl acetate)

The following procedure is typical. A solution of PSt macro RAFT agent **2-PSt** (0.030 g, 0.01 M) (Table 9, Expt 5; M_n 1850), AIBN (1.4 mg, 0.005 M) and VAc (1.4 g) was transferred to an ampoule which was degassed by three freeze-evacuate-thaw cycles and sealed. The ampoule was heated at 70 °C for the times indicated in Table 10. The volatiles were removed *in vacuo* to give PSt-*block*-PVAc. ${}^{NMR}M_n$ was determined from the ratio of regions δ 4.6-5.2 (PVAc O_2CCH_3) and δ 6.2-7.4 (PSt Ar-H) in the ^1H NMR spectra.

Preparation of Polystyrene-block-Poly(methyl acrylate)

A solution consisting of **2-PSt** (0.56 g) (Table 9, Expt 3; M_n 10,500, M_w/M_n 1.06), AIBN (8.2 mg), MA (2.5 mL) and benzene (2 mL, total volume 5 mL) was prepared in an ampoule which was degassed by three freeze-evacuate-thaw cycles and sealed. The ampoule was heated at 70 °C for 3 hours. The volatiles were removed *in vacuo* to give the PSt-*block*-PMA macroRAFT agent (**2-PSt-PMA-H$^+$**) at 1.15 g (24.6%), with M_n 21,500, M_w/M_n 1.15. The ${}^{NMR}M_n$ 22,000 was calculated considering the regions δ 3.2-4.0 ppm (PMA OCH_3) and 6.2-7.4 ppm PSt Ar-H) of the ^1H NMR spectra. The residue was dissolved in dichloromethane and percolated through a carefully crushed and dried sodium carbonate bed. The color of the solution changed from yellow to colourless. Removal of the solvent gave **2-PSt-PMA** macroRAFT agent.

Preparation of Low Dispersity Poly(styrene)-block-Poly(methyl acrylate)-block-Poly(vinyl acetate)

A solution consisting of the **2-PSt-PMA** (0.5 g) ($^{NMR}M_n$ 21,500, M_w/M_n 1.15), AIBN (1.64 mg) and VAc (1.65 mL) was prepared in an ampoule which was degassed by three freeze-evacuate-thaw cycles and sealed. The ampoule was heated at 70 °C for 15 hours. The volatiles were removed *in vacuo* to give PSt-*block*-PMA-*block*-PVAc (**2-PSt-PMA-PVAc**) at 1.60 g (73.6%), with M_n 52,000, M_w/M_n 1.48. The $^{NMR}M_n$ 72,000 was calculated considering the NMR regions 4.6-5.2 ppm (PVAc O_2CCH_3) and 6.2-7.4 ppm (PSt Ar-H).

Preparation of Polystyrene-block-Poly(methyl acrylate-grad-vinyl acetate)

The following procedure is typical. A solution of **2-PSt-H$^+$** (M_n 1850, 13.5 mg, 0.01 M), AIBN (0.12 mg, 0.005 M), VAc (605 mg) and MA (32 mg) was transferred to an ampoule which was degassed by three freeze-evacuate-thaw cycles and sealed. The ampoule was heated at 70 °C for the times indicated in Table 3. The volatiles were removed *in vacuo* to give PSt-*block*-P(MA-*grad*-VAc). Results are presented in Table 3 and Figure 6.

A similar polymerization was carried out in the heated probe of the NMR spectrometer (400 MHz). The data from this experiment are also shown in Figure 6.

Molecular Orbital Calculations

Semi-empirical molecular orbital calculations (AM1 method) were performed using CS MOPAC as implemented within the ChemBio3D Ultra package (version 11.0) on structures created using ChemBioDraw Ultra (vesion 11.0) and a full geometry minimization was performed.

NMR Study of Reversibility of Protonation

Methyl 2-(methyl(pyridin-4-yl)carbamothioylthio)propanoate (**2**) (13.5 mg, 0.05 mmole) was dissolved in chloroform-d_3 (0.5 mL) to yield a very pale yellow solution which was placed in an NMR tube and the ^{1}H NMR spectrum was recorded (Figure 1a). An equimolar quantity of TsOH (9.5mg, 0.05 mmol) was added to the NMR tube, an intense golden color coloration developed immediately. The ^{1}H NMR spectrum is shown in Figure 1b (recorded ca 2.5 h after addition of acid). An equimolar quantity of 4-(*N,N*-dimethylamino)pyridine (6.1 mg, 0.05 mmol) was then added to the NMR tube. The color intensity reduced immediately. The ^{1}H NMR spectrum is shown in Figure 1c. The solution from the NMR tube was transferred to a glass vial and extracted with water (0.5 mL), the aqueous layer was discarded and the organic layer transferred to an NMR tube. The ^{1}H NMR spectrum is shown in Figure 1d.

compound **2**: ^{1}H NMR ($CDCl_3$) δ 1.53 (d, J=7.4 Hz, 3H, CHC**H**$_3$); 3.74 (s, 6H, COOC**H**$_3$ and N-C**H**$_3$); 4.64 (q, J=7.4 Hz, 1H, C**H**CH$_3$); 7.37 (br. d, J=4.3 Hz, 2H, *m*-Ar**H**); 8.75 (br s, 2H, *o*-Ar**H**). (The difference in Chemical shift for the *m*-Ar**H** and the different appearance of *o*-Ar**H** in (a) and (d) is attributed to a solvent induced shift ($CHCl_3$ vs 4-dimethylaminopyridine neutralized, water saturated $CHCl_3$).

compound **2**-H$^+$ (TsOH): ^{1}H NMR ($CDCl_3$) δ 1.59 (d, J=7.4 Hz, 3H, CHC**H**$_3$); 2.33 (s, 3H, 4-toluenesulfonate C**H**$_3$); 3.72, 3.75 (s, 6H, COOC**H**$_3$ and N-C**H**$_3$); 4.57 (q, J=7.4 Hz, 1H, C**H**CH$_3$); 7.15 (d, J=8.0 Hz, 2H, 4-toluenesulfonate *m*-Ar**H**); 7.14 (d, J=8.0 Hz, 2H, 4-toluenesulfonate *o*-Ar**H**); 7.90 (d, J=6.4 Hz, 2H, *m*-Ar**H**); 8.94 (d, J=6.4 Hz, 2H, *o*-Ar**H**).

References

1. Jenkins, A. D.; Jones, R. I.; Moad, G. *Pure Appl. Chem.* **2010**, *82*, 483–491.
2. Moad, G.; Rizzardo, E.; Thang, S. H. *Acc. Chem. Res.* **2008**, *41*, 1133–1142.
3. Moad, G.; Rizzardo, E.; Thang, S. H. *Polymer* **2008**, *49*, 1079–1131.
4. Moad, G.; Rizzardo, E.; Thang, S. H. *Aust. J. Chem.* **2005**, *58*, 379–410.
5. Moad, G.; Rizzardo, E.; Thang, S. H. *Aust. J. Chem.* **2006**, *59*, 669–692.
6. Moad, G.; Rizzardo, E.; Thang, S. H. *Aust. J. Chem.* **2009**, *62*, 1402–1472.
7. Abbreviations: LAM - less activated monomer, MA - methyl acrylate, MAM - more activated monomer, NVP - N-vinyl pyrrolidone, St - styrene, VAc - vinyl acetate, PMA - poly(methyl acrylate), PMMA - (polymethyl methacrylate), PNVP - poly(N-vinyl pyrrolidone), PSt - polystyrene, PVAc - poly(vinyl acetate), RAFT - reversible addition-fragmentation chain transfer.
8. Chiefari, J.; Mayadunne, R. T. A.; Moad, C. L.; Moad, G.; Rizzardo, E.; Postma, A.; Skidmore, M. A.; Thang, S. H. *Macromolecules* **2003**, *36*, 2273–2283.
9. Petruczok, C. D.; Barlow, R. F.; Shipp, D. A. *J. Polym. Sci., Part A: Polym. Chem.* **2008**, *46*, 7200–7206.
10. Hussain, H.; Tan, B. H.; Gudipati, C. S.; Liu, Y.; He, C. B.; Davis, T. P. *J. Polym. Sci., Part A: Polym. Chem.* **2008**, *46*, 5604–5615.
11. Nicolay, R.; Kwak, Y.; Matyjaszewski, K. *Chem. Commun.* **2008**, 5336–5338.
12. Tong, Y.-Y.; Dong, Y.-Q.; Du, F.-S.; Li, Z.-C. *Macromolecules* **2008**, *41*, 7339–7346.
13. Huang, C.-F.; Nicolaÿ, R.; Kwak, Y.; Chang, F.-C.; Matyjaszewski, K. *Macromolecules* **2009**, *42*, 8198–8210.
14. Ray, B.; Kotani, M.; Yamago, S. *Macromolecules* **2006**, *39*, 5259–5265.
15. Debuigne, A.; Caille, J.-R. l.; Willet, N.; Jerome, R. *Macromolecules* **2005**, *38*, 9488–9496.
16. Quemener, D.; Davis, T. P.; Barner-Kowollik, C.; Stenzel, M. H. *Chem. Commun.* **2006**, 5051–5053.
17. Xue, X.; Zhu, J.; Zhang, Z.; Cheng, Z.; Tu, Y.; Zhu, X. *Polymer* **2010**, *51*, 3083–3090.

18. Benaglia, M.; Chiefari, J.; Chong, Y. K.; Moad, G.; Rizzardo, E.; Thang, S. H. *J. Am. Chem. Soc.* **2009**, *131*, 6914–6915.
19. Benaglia, M.; Chen, M.; Chong, Y. K.; Moad, G.; Rizzardo, E.; Thang, S. H. *Macromolecules* **2009**, *42*, 9384–9386.
20. Rizzardo, E.; Chiefari, J.; Mayadunne, R. T. A.; Moad, G.; Thang, S. H. *ACS Symp. Ser.* **2000**, *768*, 278–96.
21. Mayadunne, R. T. A.; Rizzardo, E.; Chiefari, J.; Chong, Y. K.; Moad, G.; Thang, S. H. *Macromolecules* **1999**, *32*, 6977–6980.
22. Thang, S. H.; Chong, Y. K.; Mayadunne, R. T. A.; Moad, G.; Rizzardo, E. *Tetrahedron Lett.* **1999**, *40*, 2435–2438.
23. Moad, G.; Solomon, D. H. *The Chemistry of Radical Polymerization*, 2nd ed.; Elsevier: Oxford, 2006; p 269.
24. Fischer, H.; Radom, L. *Angew. Chem., Int. Ed.* **2001**, *40*, 1340–1371.
25. Kaneyoshi, H.; Matyjaszewski, K. *Macromolecules* **2005**, *38*, 8163–8169.
26. Koumura, K.; Satoh, K.; Kamigaito, M. *J. Polym. Sci., Part A: Polym. Chem.* **2009**, *47*, 1343–1353.
27. Bicciocchi, E.; Chong, Y. K.; Giorgini, L.; Moad, G.; Rizzardo, E.; Thang, S. H. *Macromol. Chem. Phys.* **2010**, *211*, 529–538.

Chapter 8

Asymmetric Micellization of Oragnometallic Polyether Block Copolymers

Christine Mangold,[1] Frederik Wurm,[2,*] and Andreas F. M. Kilbinger[3]

[1]Institute of Organic Chemistry, Organic and Macromolecular Chemistry, Duesbergweg 10-14, Johannes Gutenberg-Universität Mainz, D-55099 Mainz, Germany
[2]Institut des Matériaux, Laboratoire des Polymères, Batiment MXD, Station 12, Ecole Polytechnique Fédérale de Lausanne, CH-1015 Lausanne, Switzerland
[3]University of Fribourg, Department of Chemistry, Chemin du Musée 9, CH-1700 Fribourg, Switzerland
***frederik.wurm@epfl.ch**

Anionic ring-opening polymerization was applied for the synthesis of several di- and triblock copolymers bearing a variable amount of vinyl-ether moieties. These vinyl ethers were reacted with Grubbs' catalyst to generate an organometallic copolymer via a stable Fischer carbene. The block copolymers were investigated via TEM for their aggregation behavior.

Introduction

The synthesis of block copolymers can be regarded as one of the key-features of macromolecular science. Block copolymers offer a tremendous field for applications mostly due to the possibility of combining different characteristics of the respective homopolymers, but also due to incompatibility of both segments resulting in a great variety of supramnolecular aggregates (*1*). Within the field of block copolymers, linear diblock copolymers are the most investigated class and their supramolecular structures in solution and in bulk have been studied in detail and are well-understood. Research in this field covers not only coiled block copolymers, but also other types of block copolymers like rod-coil block copolymers (*2–4*) or nonlinear block copolymers, i.e. stars (*5*) or highly branched segments (*6*). Introducing a third block results in terpolymers and the variety of possible superstructures increases even further (*7*). Until now there are only few literature examples focussing on the aggregation behaviour of block terpolymers in comparison to the abundance of reports investigating diblock copolymers.

In the last decades the synthesis of asymmetric nanoparticles, which are often called "Janus-particles", have gathered increased attention. Since de Gennes' Nobel lecture in 1991, in which he presented the term "Janus grains" (*8*), different areas of researchers have focussed on the synthesis of asymmetric particles. Macroscopic, microscopic and nanoscopic particles have been prepared in which certain parts of their surface differ in chemical composition, polarity, color, or any other property. Spherical, cylindrical, disc-like, snowman-, hamburger-, and raspberry-like structures have been synthesized from organic as well as inorganic materials, or even as hybrids of both. The term "Janus" originates from the two-faced Roman god Janus (Figure 1), the god of the doors. Even in modern culture, the month of January, the first month of the New Year, and the janitor, who is a caretaker of doors and halls, remind us of this ancient Roman god. Janus is depicted as a double-headed god who represents dichotomy. Some recent reviews give an overview on different Janus particles in detail (*9–11*).

Within this field of research, linear and miktoarm ABC-block terpolymers have been used for the construction of multicompartment micelles, which are also referred to as janus micelles. Linear ABC-block terpolymers have been ordered in the bulk or on solid surfaces to prepare janus micelles, which are asymmetric star copolymers, after physical cross-linking. Since the first mention of janus micelles, janus discs and janus cylinders have also been reported and generated increased academic as well as technological interest (*9*, *12*).

We recently presented the first Janus particles which are based on an amphipjilic polyether diblock copolymer undergoing a cross metathesis reaction with Grubbs' first generation catalyst (*13*).

In addition to the unexpected formation of asymmetric aggregates this technique allows TEM staining of polymers by the attachment of the electron rich ruthenium to the "invisible" polyether chain. A principal problem in analyzing supramolecular polymeric or compartmented structures is their visualization. Typically, scanning force microscopy (SFM) or transmission electron microscopy (TEM) can be employed. The differentiation between individual compartments or constituents of the supramolecular assembly can be particularly difficult or impossible. In many cases where transmission electron microscopy is used for visualization of polymeric structures, heavy metal staining agents are employed to increase the image contrast between areas of different chemical composition (*14*). The most common examples are osmium tetroxide which has been widely employed as a stain for unsaturated polymers, ruthenium tetroxide for saturated and unsaturated polymers or phosphotungstic acid which is especially useful for polyamide staining (*14*).

To our surprise, the commercially available ruthenium carbene catalysts reported by Grubbs et al. (*15–17*) have to date not been employed in TEM staining of polymers. Compared to the highly toxic and volatile osmium tetroxide they offer lower toxicity and easier handling at lower cost.

Figure 1. Typical Roman statue of the god Janus and different Janus particles that have been reported in literature (taken from reference (9)).

This chapter will summarize the results for the previously mentioned diblock copolymers based on poly(ethylene oxide) (PEO) and poly(propylene oxide) (PPO) (*13*) and will introduce a novel block copolymer with a highly hydrophobic block based on glycidyl-(1,1,2,2-tetrafluoro-ethyl)-ether. In addition we will present some first results on ABC-block terpolymer systems with a variable number of organometallic moieties attached to a polyether backbone via cross metathesis (*18*).

Results and Discussion

Diblock Copolymers

All polymers presented herein rely on classical anionic ring-opening polymerization (ROP) of epoxides initiated by cesium alkoxides.

We used the cesium salt of 2-(vinyloxy)ethanol (**2**) for the synthesis of diblock copolymers based on PEO and PPO via anionic polymerization bearing a single vinylether at the polyether chain end, either at the hydrophobic (PPO) (**5**) or hydrophilic (PEO) (**3**) end by variation of the monomer addition sequence. These narrowly distributed block copolymers were subsequently modified with a ruthenium carbene comlex, namely the first generation Grubbs catalyst (**1**) to introduce an organometallic moiety at the chain end via a stable Fischer carbene linkage. It has been well established that vinyl ethers react with ruthenium carbene catalysts to form Fischer carbene-type complexes that are effectively inactive for further olefin metathesis reactions (*19*). Reactions of this type are therefore often used for the non-functional chain termination of ring opening metathesis polymerizations (ROMP).

We also synthesized a hitherto unknown block copolymer, poly(ethylene oxide)-*block*-poly(glycidyl 1,1,2,2-tetrafluoro-ethyl ether) (**4**), again carrying a vinyl ether moiety at the hydrophilic chain end. The molecular weights of the narrow polydispersity block copolymers ranged from ca. 4,200-6,000 g mol^{-1} (Table 1).

The synthesis of the diblock copolymers and their modification are shown in Scheme 1. Their supramolecular aggregation in water was monitored via

Table 1. Molecular weight data for diblock copolymers

#	M_n *(SEC)*[a]	*PDI*[a]
3 Vinyl-PEO$_{102}$-PPO$_{15}$	5,300	1.06
4 Vinyl-PEO$_{96}$- PTFGE$_{10}$	6,000	1.07
5 Vinyl-PPO$_{13}$-PEO$_{74}$	4,200	1.05

[a] determined via SEC in DMF vs PEO Standards.

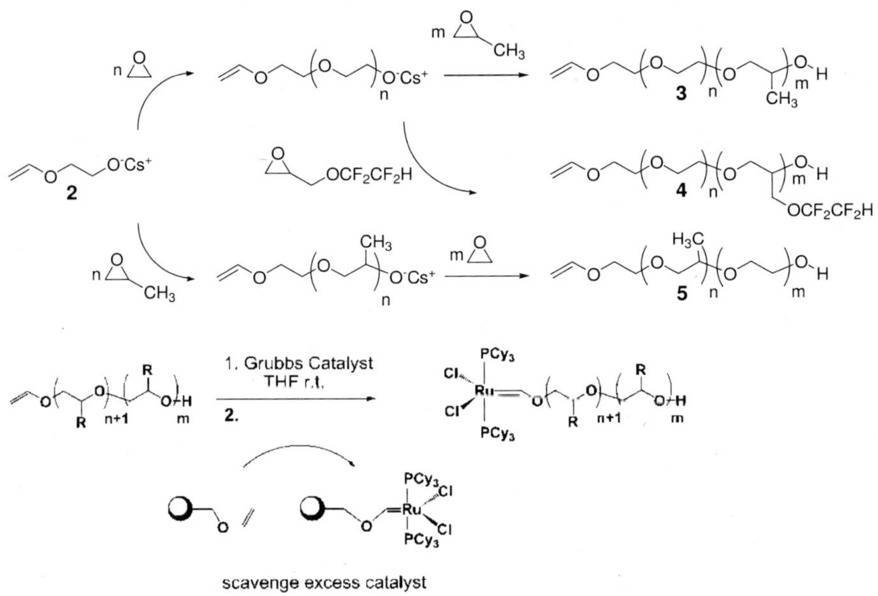

Scheme 1. Top: Synthesis of diblock copolymers. Bottom: General procedure for cross metathesis of vinyl ether group(s) with Grubbs catalyst.

transmission electron microscopy. A high contrast of the micelles was expected due to the presence of the electron-rich ruthenium in the micelle.

Solutions of unmodified polymers **3**, **4** and **5** were prepared in THF (1 mg mL^{-1}), which is a good solvent for all segements. They were then further diluted with water to induce micellation. Due to lack of contrast the micelles could be observed using TEM.

Addition of an excess of **1** to the THF solutions of **3**, **4** and **5** resulted in an olefin metathesis reaction at the chain end of the block copolymers. This leads to the formation of a Fischer-type carbene, which is indicated by the color change from pruple to light brown. This reaction sequence allows the covalent attachment of the Grubbs catalyst to the block copolymer chain end (Scheme 1). After dilution with water and removal of the insoluble excess of **1**, TEM images were recorded of all three solutions, which revealed dark spherical objects with enhanced contrast compared to the TEM grid (not shown). We believe that parts of the ruthenium catalyst **1** were dissolved in the hydrophobic core of the formed micelles, leading to non-specific staining.

To remove virtually all excess of catalyst **1**, we prepared a scavenger resin which was functionalized with vinyl ether groups (*13*). In a subsequent experiment we added an excess of **1** to THF solutions of **3**, **4** and **5**, followed by the removal of the excess by the addition of the scavenger resin. For polymer **5** dark spherical micelles were observed, as the organometallic group and the PPO-block form the the micellar core, while PEO solubilizes the aggregate in the corona (Figure 2, A). For polymers **3** and **4** interestingly non-symmetrical aggregates (Figure 2, B, C) were observed which coexisted with supermicelles (Figure 2, D,E,F) as reported for other Janus micelles (for detailed information on aggregation model see reference (*13*)).

To further investigate the micellization and staining process, we modified the terminal vinyl-bond by the addition of bromine and visualized the micelles from the aqueous solution via TEM. As expected, all diblock copolymers, regardless if modified on the hydrophilic or hydrophobic chain end, aggregate to conventional spherical micelles with a slightly increased TEM-contrast, which is due to the presence of the bromides at the chain end (Figure 3).

This led us to the conclusion that cross metathesis between the terminal vinyl group and the carbene catalyst, does not only enhance the TEM contrast but in addition has a strong influence on the aggregation behavior of the materials. It has to be pointed out that the strongly hydrophobic character and the quite high molecular weight of the Grubbs catalyst (ca. 800 g/mol) generate an ABC-system similar to an ABC-terpolymer with the organometallic unit representing a segment of similar molecular weight as the hydrophobic PPO-block. By decreasing the molecular weight of the “covalent stain” to bromine, conventional micellization was observed with a slight staining effect.

Terblock Copolymers

In a following step the findings from the abovementioned diblock copolymer systems bearing only one organometallic side chain were extended to a block terpolymer system with a variable number of vinyl groups for attachment of the Fischer carbene. For this purpose we designed a novel monomer for anionic ROP, i.e. ethoxy vinyl glycidyl ether (EVGE). The monomer can be conveniently synthesized from epichlorohydrin and 2-(vinyloxy)ethanol under phase-transfer conditions as reported for other glycidyl ethers before (*20*, *21*) (Scheme 2) and was isolated by fractionated distillation.

ABC-terpolymers were then synthesized with propylene oxide as the first, ethylene oxide as the second, and EVGE as the last block by sequential anionic ROP. By the degree of polymerization of EVGE the number of possible metathesis sites can be adjusted. Scheme 3 shows the synthetic strategy and Figure 4 shows the SEC elugrams of the block copolymer after each block.

As it can be clearly seen narrow molecular weight distributions are obtained for all blocks. Table 2 summarizes the molecular weight data. Figure 5 shows the ^{1}H NMR spectra for all stages of the polymerization from which the degree of polymerization of the terpolymer can be calculated accurately by comparison to the methyl group of the initiator (in this case methoxy ethanol). Furthermore the ^{1}H NMR spectra prove the successful incorporation of EVGE

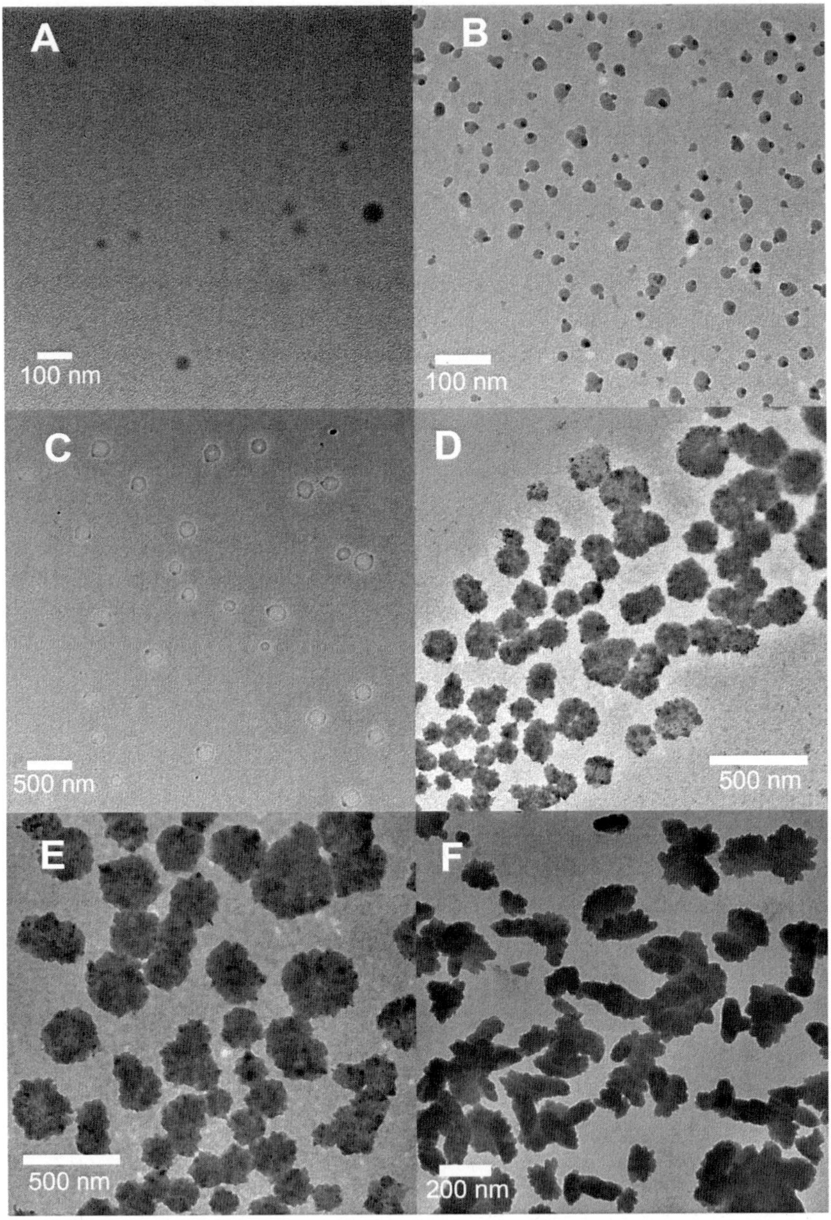

*Figure 2. TEM images of block copolymers **3,4,5** reacted with Grubbs catalyst 1st generation dropcast from an aqueous solution. A:**5-Ru**. B:**3-Ru**. C:**4-Ru**. D: supermicelles of 3-Ru E: zoom into supermicelles of **3-Ru** F: supermicelles **4-Ru**.*

with retention of its vinyl-functionality. The respective resonances for the vinyl protons appear at ca. 6.5 ppm and between 4.1 and 4.0 ppm. In addition to the non-functional methoxyethanol which was applied as the initiator, we also applied *N,N*-dibenzylaminoethanol as an initiator (*22*) to introduce an

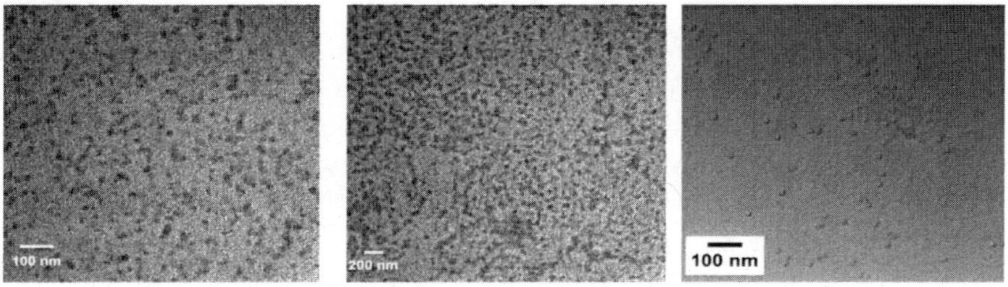

Figure 3. TEM pictures of brominated diblock copolymers left: ***Br-3****, middle:* ***Br-5****, right:* ***Br-4****.*

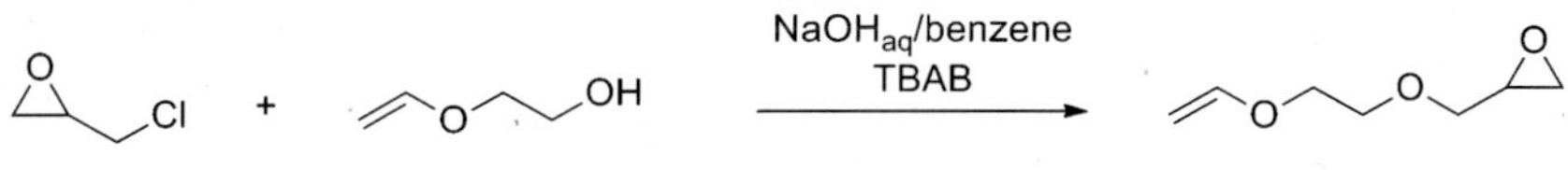

Scheme 2. Synthesis of ethoxy vinyl glycidyl ether under phase transfer conditions (TBAB= tetrabutylammonium bromide).

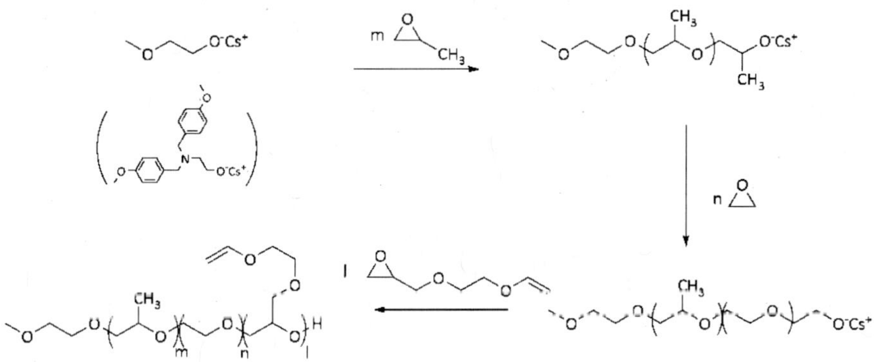

Scheme 3. Synthesis of ABC-triblock terpolymers bearing several vinyl ether groups along the backbone.

additional aminofunctionality allowing orthogonal reactivity (not presented in this manuscript).

MALDI-ToF MS further confirmed the successful incorporation of all monomers into the polymer as the distances for all monomer repeat units can be found in the spectrum. Figure 6 shows a zoomed-in section of a typical mass spectrum of an ABC-terpolymer, highlighting its complex features due to the presence of three distinct distributions.

In contrast to the AB-linear block copolymers, with only one Grubbs catalyst attached at the chain-end, the aggregation behavior of the triblock terpolymers as investigated by TEM was found to be more complex. A similiar staining procedure was applied as reported above, i.e. addition of an excess of **1** and subsequent treatment with the vinylether-modified scavenger resin. In contrast to the abovementioned process for the diblock copolymers, the formation of the Fischer carbene, showed slow kinectics, which were investigated in detail by time-resolved ^{1}H NMR experiments (data not shown here); thus, the reaction

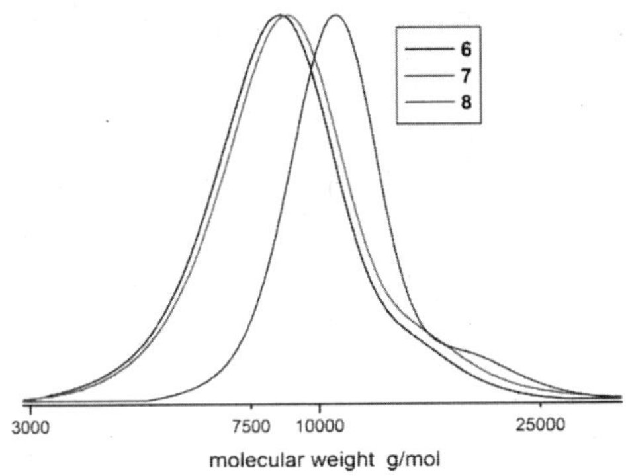

Figure 4. Molecular weight distribution of samples ***6-8*** *in THF. Additional data is given in Table 2.*

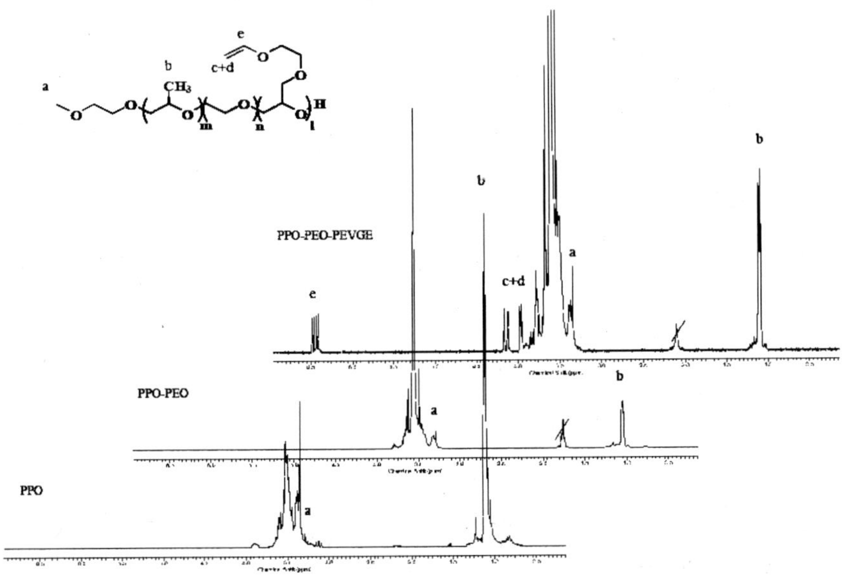

Figure 5. ^{1}H *NMR spectra (CDCl3, 300 MHz) of samples taken after polymerization of each block; lower spectrum: PPO-OH, middle: PPO-b-PEO-OH, top: PPO-b-PEO-b-PEVGE-OH.*

time was extended to several hours to guarantee a nearly quantitative (up to 90%) conversion of all vinyl-ether moieties.

Copolymer **6**, whose microstructure is close to copolymer **3,** forms similar aggregates, as can be seen in Figure 7. This finding is consistent with theoretical expectations. The vinyl-units of the terpolymers **7** and **8** were also reacted to the corresponding organometallic polymers, but did not exhibit reproducible structures in TEM. This might be ascribed to the very high molecular weight of the organometallic block, which seems to suppress the influence of the two other blocks. We found that our previous procedure used to visualize such structures in

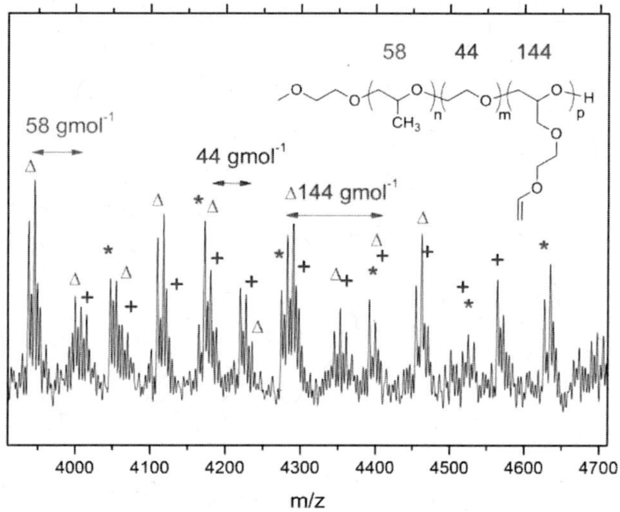

Figure 6. Zoom into a MALDI-ToF spectrum of the terpolymer PPO-b-PEO-b-PEVGE (measured with CHCA as matrix and KTFA as a cationizing agent; triangles represent the monomer distance for PO (58 g/mol), stars the distance for EVGE (144 g/mol) and the crosses represent the EO distance of 44 g/mol).

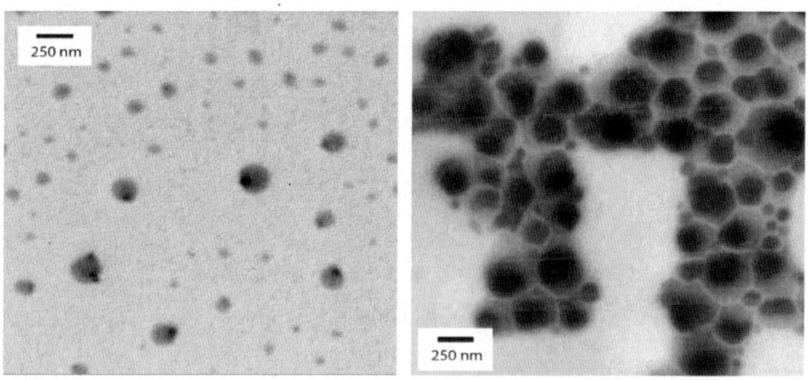

*Figure 7. TEM images of block copolymer **6** reacted with Grubbs' 1st generation catalyst dropcast from an aqueous solution. Left: unimers which coexist with supermicelles (right).*

TEM is not suitable for these terpolymers as the water-solublity after modification is low. Ongoing studies investigate the variation of block ratios to obtain reliable results. Nevertheless, this first example proves the versatility of this approach to generate novel organometallic polymers with a nonconventional aggregation into asymmetric micelles, induced by olefin metathesis. In addition, by using EVGE as comonomer, one is still able to introduce an orthogonal by the initiator, e.g. amines or orther chelating groups could be possible.

Table 2. Molecular weight data of ABC-terpolymers

#	M_n *(NMR)*[a]	*PDI*[b]
6 PPO_{15}-*b*-PEO_{156}-*b*-$PEVGE_{1.5}$	7,800	1.11
7 PPO_{15}-*b*-PEO_{156}-*b*-$PEVGE_{6}$	8,500	1.09
8 PPO_{15}-*b*-PEO_{156}-*b*-$PEVGE_{16}$	11,800	1.09[c]

[a] determined from ^1H NMR by comparison to the initiator protons. [b] Determined via SEC in THF vs. PEO-standards. [c] Showed a slightly bimodal distribution.

Conclusions

In summary, we have developed a versatile method for polymer staining using a ruthenium based irreversible olefin metathesis reaction which can be triggered along the polymer backbone by either initiating with a vinyl ether-based alcohol or by incorporation of a vinylether containing monomer into the polymer structure. Efficient removal of any excess of the Grubbs catalyst was performed by a specially designed scavenger resin, making specific staining visible. This scavenger resin could be also applied as a convenient protocol for removing the carbene catalyst from conventional metathesis reactions to prevent other purification steps such as chromatography.

The organometallic polyether block copolymers exhibit interesting supramolecular structures which are strongly influenced by the attached ruthenium moieties. In addition, first results of novel triblock terpolymers with variable amount of vinyl were presented.

While further and more detailed studies are currently undertaken, we assume that for larger polymers the effect on new architectures is negligible and pure staining can be observed due to the high contrast of ruthenium in TEM. This observation renders the described method highly versatile and applicable to many other materials, avoiding very toxic or radioactive reagents relying on well-established olefin metathesis.

Experimental

Instrumentation

^1H NMR spectra (300 MHz) were recorded using a Bruker AC300. All spectra were referenced internally to residual proton signals of the deuterated solvent. Size exclusion chromatography (SEC) measurements were carried out in THF on an instrument consisting of a Waters 717 plus autosampler, a TSP Spectra Series P 100 pump, a set of three PSS SDV columns (104/500/50Å), and RI and UV detectors. All SEC diagrams rely on the RI detector signal, and the molecular weights refer to linear poly ethylene oxide (PEO) standards provided by Polymer Standards Service. Matrix-assisted laser desorption and ionization time-of-flight (MALDI-ToF) measurements were performed on a Shimadzu Axima CFR MALDI-TOF mass spectrometer equipped with a nitrogen laser delivering 3ns laser pulses at 337nm. α-cyano-4-hydroxy cinnamic acid (CHCA) was used as matrix. Samples were prepared by dissolving the polymer in acetonitrile at a concentration of 1g/L.

A 10μL aliquot of this solution was added to 10μL of a 10g/L solution of the matrix and 1μL of a solution of KTFA (0.1M in water as cationization agent). A 1μL aliquot of the mixture was applied to a multistage target and allowed to dry to create a thin matrix/analyte film. The samples were measured in positive ion and in linear or reflection mode of the spectrometer. A Philips EM 420 transmission electron microscope using a LaB6 cathode at an acceleration voltage of 120 kV was used to obtain TEM-images. TEM grids (carbon film on copper, 300 mesh) were obtained from Electron Microscopy Sciences, Hatfield, PA, USA.

Reagents

All solvents and reagents were purchased from Acros Organics and used as received, unless otherwise stated. Chloroform-d_1 was purchased from Deutero GmbH. Polymerization procedure for the diblock copolymers based on PEO and PPO and scavenger resin can be found elsewhere (*13*).

Ethoxy Vinyl Glycidyl Ether (EVGE)

2-(vinyloxy)ethanol (10 g, 113.5 mmol) was placed in a 500 mL round bottom flask and dissolved in a mixture of 50% aqueous NaOH (150 mL) and benzene (150 mL). To this mixture tetrabutylammonium bromide (3.5 g, 11 mmol) was added and the mixture was stirred quickly with a mechanical stirrer. Then the reaction mixture was cooled with an ice bath and epichlorohydrin (31.5 g, 340.5 mmol) was slowly added via a dropping funnel. After 24h reaction time at room temperature, the organic phase was separated from the aqueous phase, washed several times with brine, dried and concentrated *in vacuo* to remove benzene and the excess epichlorohydrin. The resulting slightly yellow residue was distilled under reduced pressure to yield the desired product as a colorless liquid in typically in 70-80% yield (11-13 g). H NMR ($CDCl_3$, 300 MHz): δ= 6.44 (dd, CH_2=C*H*, J_1=14.3, J_2=7), 4.13 (dd, *CH₂*=CH, J_1=14.3, J_2=2.2), 3.96 (dd, CH_2=CH, J_1=7, J_2=2.2), 3.8-3.65 (m, -O-CH_2-CH_2-O- 4H & CH_2 (*glycidyl ether*) 1H), 3.38 (dd, CH_2 (*glycidyl ether*), 1 H, J_1=11.8, J_2=5.9), 3.1 (m, C*H-epoxide*, 1H), 2.74 (dd, CH_2*-epoxide*, 1H, J_1=5, J_2=4.2), 2.56 (dd, CH_2*-epoxide*, 1H, J_1=5.2, J_2=2.6).

General Procedure for the Synthesis of the Terpolymers P(PO-*block*-EO-*block*-EVGE)

The corresponding initiator was dissolved in benzene in a 250 mL-Schlenk flask and 0.9 equivalents of cesium hydroxide were added. The mixture was stirred at 60 °C under argon for 1 h and evacuated at (10^{-2} mbar) for 6 h to remove benzene and water, forming the corresponding cesium alkoxide. Propylene oxide was first cryo transferred into a graduated ampoule and then into the Schlenck flask. The polymerization was carried out at 60°C. After 24h approximately 15 mL dry THF was cryo-transferred into the Schlenk flask. Ethylene oxide was first cryo-transferred to a graduated ampoule, and then cryo-transferred into the flask containing PO-Oligomers. The mixture was again heated to 60 °C and stirred for 12h. Before the addition of EVGE, it was dried with CaH_2 and distilled freshly.

The third monomer was added using a syringe and the solution was heated to 90°C for additional 12h. Precipitation in cold diethyl ether resulted in the pure copolymers. Yields: 95% to quantitative. ^{1}H NMR (300 MHz, $CDCl_3$): δ= 6.46 (dd, $H_2C{=}CH$), 4.15 (dd, $HHC{=}CH$), 3.97 (dd, $HHC{=}CH$), 3.68-3.34 (polyether backbone) 1.15 (m, CH_3). Additional signals occur in dependence of initiator: $(MeOBn)_2NC_2H_4OH$: 7.21 (d, 2H, arom.) 6.81 (d, 2H, arom.), 3.76 (s, 6H, OCH_3), 3.60 (s, 4H, NCH_2Ph), 3.57 (t, 2H,CH_2OH) 2.60 (t, 2H, NCH_2CH_2O-).

Staining and Micellization via Grubbs' Catalyst

For staining with the 1st generation Grubbs' catalyst, the calculated amount of polymer (ca. 5 mg) was dissolved in degassed THF (concentration ca. 50 mg mL-1) and 1.1 eq. Grubbs' catalyst was added. The mixture was placed on an orbital shaker for 2 to 4 hours.

Without a scavenger resin: After the desired reaction time, degassed water (5 mL) was added to the polymer solution it was allowed to equilibrate at room temperature for 1 h and then ca. 2 μL were transferred on a TEM-copper grid and dried over night in vacuo.

With scavenger resin: The mixture was diluted with THF to approx. 10 mg mL-1, the calculated amount of scavenger resin (2 eq excess, at theor. 100% degree of functionalization) was added and the mixture was placed on an orbital shaker for an additional 2-4 hours. Finally, the resin was removed by filtration and the solution was concentrated again to approx. 50 mg mL-1, diluted with water (100-200 μL of THF solution was diluted with 5 mL of water), equilibrated for 0.5 - 3h at room temperature and then drop cast onto a TEM grid and dried over night *in vacuo*.

Acknowledgments

The authors thank Prof. Holger Frey for his support. A.F.M.K. acknowledges the "Fonds der chemischen Industrie" for valuable financial support. F.W. is grateful to the Alexander von Humboldt-Stiftung for Feodor-Lynen research fellowship.

References

1. Bates, F. S.; Fredrickson, G. H. *Phys. Today* **1999**, *52*, 32–38.
2. Lee, M.; Cho, B.-K.; Zin, W.-C. *Chem. Rev.* **2001**, *101*, 3869–3892.
3. Olsen, B. D.; Segalman, R. A. *Mater. Sci. Eng. R: Reports* **2008**, *62*, 37–66.
4. König, H.; Kilbinger, A. *Angew. Chem., Int. Ed.* **2007**, *46*, 8334–8340.
5. Hadjichristidis, N. *J. Polym. Sci., Part A: Polym. Chem.* **1999**, *37*, 857–871.
6. Wurm, F.; Frey, H. *Prog. Polym. Sci.*, in press, Corrected Proof.
7. Hadjichristidis, N.; Iatrou, H.; Pitsikalis, M.; Pispas, S.; Avgeropoulos, A. *Prog. Polym. Sci.* **2005**, *30*, 725–782.
8. De Gennes, P. G. *Rev. Mod. Phys.* **1992**, *64*, 645.
9. Wurm, F.; Kilbinger, A. *Angew. Chem., Int. Ed.* **2009**, *48*, 8412–8421.

10. Walther, A.; Müller, A. H. E. *Soft Matter* **2008**, *4*, 663–668.
11. Perro, A.; Reculusa, S.; Ravaine, S.; Bourgeat-Lami, E.; Duguet, E. *J. Mater. Chem.* **2005**, *15*, 3745–3760.
12. Nie, L.; Liu, S.; Shen, W.; Chen, D.; Jiang, M. *Angew. Chem., Int. Ed.* **2007**, *46*, 6321–6324.
13. Wurm, F.; König, H. M.; Hilf, S.; Kilbinger, A. F. M. *J. Am. Chem. Soc.* **2008**, *130*, 5876–5877.
14. Smith, R. W.; Bryg, V. *Rubber Chem. Technol.* **2006**, *79*, 520–540.
15. Nguyen, S. T.; Grubbs, R. H.; Ziller, J. W. *J. Am. Chem. Soc.* **1993**, *115*, 9858–9859.
16. Schwab, P.; Grubbs, R. H.; Ziller, J. W. *J. Am. Chem. Soc.* **1996**, *118*, 100–110.
17. Nguyen, S. T.; Johnson, L. K.; Grubbs, R. H.; Ziller, J. W. *J. Am. Chem. Soc.* **1992**, *114*, 3974–3975.
18. Wurm, F. *Polym. Prepr. (Am. Chem. Soc., Div. Polym. Chem.)* **2010**, *51*, 322–323.
19. Grubbs, R. H.; Chang, S. *Tetrahedron* **1998**, *54*, 4413–4450.
20. Wurm, F.; Nieberle, J.; Frey, H. *Macromolecules* **2008**, *41*, 1909–1911.
21. Obermeier, B.; Wurm, F.; Frey, H. *Macromolecules* **2010**, *43*, 2244–2251.
22. Wurm, F.; Klos, J.; Räder, H. J.; Frey, H. *J. Am. Chem. Soc.* **2009**, *131*, 7954–7955.

Chapter 9

Stimuli Responsive Triblock Copolymers – Synthesis, Characterization, and Application

John Texter,[1,*] Vivek Arjunan Vasantha,[1] Kejian Bian,[1] Xiumin Ma,[1] Lisa Slater,[2] Thomas Mourey,[2] and Gary Slater[2]

[1]Coatings Research Institute, School of Engineering Technology, Eastern Michigan University, Ypsilanti, MI 48107, USA
[2]Corporate Research and Engineering, Eastman Kodak Company, Rochester, NY 14650-2136, USA
[*]jtexter@emich.edu

Triblock copolymers of a hydrophobic polypropylene oxide (PPO) block with endblocks derived by ATRP of an ionic liquid reactive surfactant, ILBr, where the PPO block was converted into a bifunctional macroinitiator are reported. A methanol/water CuBr system shown earlier to produce ILBr homopolymers by ATRP was used to grow the ILBr blocks. Good control was not observed in polymerization, but the products exhibit high surface activity and stimuli responsiveness. These triblocks appear more surface active than comparable molecular weight polyethylene oxide-polypropylene oxide-polyethylene oxide triblocks and have lower critical micelle concentrations. They are also anion stimuli responsive and surface activity can be tuned by ion exchange. These triblocks represent a new class of steric stabilizers and can be used effectively in high salt.

Introduction

The initial report (*1*) of using ILBr as an ionic liquid monomer for making stimuli responsive gels, membranes, and latexes showed that latexes comprising ILBr did not coagulate in high salt, such as in 0.1 M NaBr. Further studies showed that such ultrastable nanolatexes could only be coagulated at relatively high concentrations of various stimuli triggering anions, such as BF_4^-, PF_6^-, and $S^=$ (*2*,

3) where the condensing anion appeared to bind strongly to the imidazolium ring, turning the hygroscopic imidazolium bromide into a much more hydrophobic ion pair. A more recent report (*4*) of making distimuli responsive copolymers of ILBr and of *N*-isopropylacrylamide illustrated the thermal condensation of PNIPAM core – polyILBr corona particles that appeared impervious to boiling and maintained their colloidal stability. These observations suggest that the imidazolium bromide moiety can provide very effective steric stabilization. It therefore seemed desirable to produce triblock copolymers comprising ILBr blocks in a series of stabilizers analogous to the well known Pluronic® series having central blocks composed of polypropylene oxide and endblocks composed of polyethylene oxide (*5*).

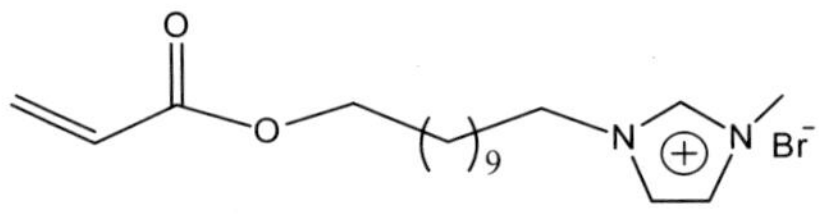

ILBr

Triblock copolymers incorporating the hydrophobic polypropylene oxide (PPO) block with endblocks derived by ATRP of the ionic liquid reactive surfactant, ILBr, were therefore designed. We report converting PPO blocks of about 1,000 and 3,500 Daltons, respectively, into macroinitiators **I** and **II**. A methanol/water CuBr system shown recently (*6*) to produce ILBr homopolymers by ATRP was used to grow the ILBr blocks. Good control was not observed, but the resulting triblocks appear quite surface active. These triblocks are also anion stimuli responsive, and surface activity can be tuned by ion exchange. We expect that these triblocks represent a new class of steric stabilizers that will prove effective in low and high salt, despite being ionic in nature.

Experimental

Materials

The following were purchased from Sigma-Aldrich: Acryloyl chloride (96%), 11-bromo-1-undecanol (98%), triethylamine (99.5%), 1-methylimidazole (99%), 2,6-di-tert-butyl-4-methylphenol (99%), neutral aluminum oxide (activated, Brockman I standard grade, ~150 mesh), poly(propylene glycol)s (PPG, M_n 1000 and 3500 respectively), 2-bromoisobutyrylbromide (BiBB, 98%), 1,1,4,7,10,10-hexamethyltriethylenetetramine (HMTETA, 97%), copper (I) bromide (99.99%), sodium bicarbonate, anhydrous sodium sulfate (≥99%) and anhydrous magnesium sulfate (≥97%) were used without further purification. Tetrahydrofuran (THF, ≥99.9%), dichloromethane (≥99.8%) and toluene (99.8%) were anhydrous and obtained from Aldrich. Diethyl ether was obtained from Fisher Scientific. Deionized water was used for all experiments. Yellow pigment PY138 and 100 μm diameter zircon grinding media were the gift of Eastman Kodak Company. The latex dispersion UCAR 300 was kindly provided by Dow Chemical.

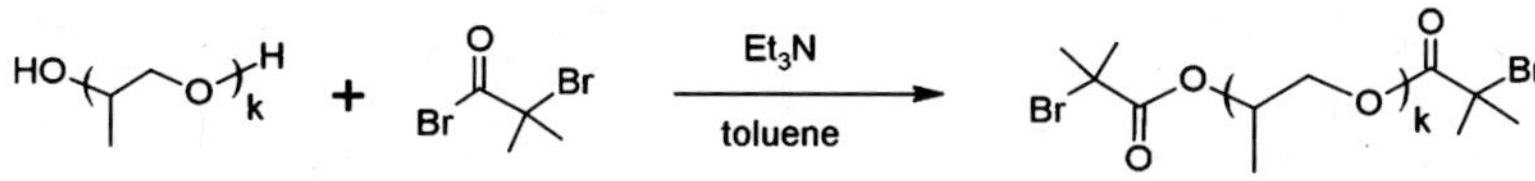

Figure 1. Scheme for preparing bifunctional macroinitiator.

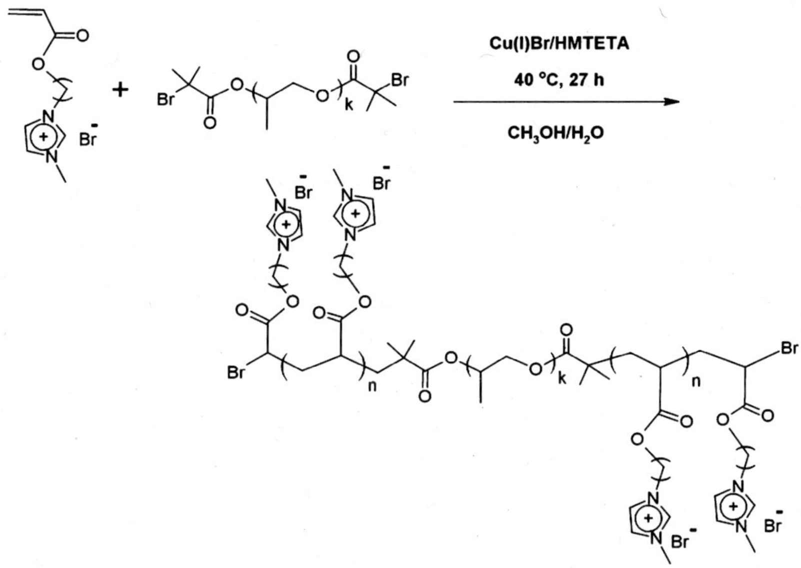

Figure 2. Scheme for synthesizing triblocks.

Instrumentation

^{1}H NMR experiments were performed on a JEOL 400 MHz NMR. A 5 mm O.D. NMR tube was used and the spectrum was accumulated with scans of 16 for macroinitiators and with 32 for all block copolymer samples. DSC analysis was performed on a TA Instruments DSC Q2000. Tzero aluminum pans and lids were used. The samples were heated at 10°C/min to 110°C and held isothermally for 5 min to remove their thermal history. They were then cooled at 5°C/min to -180°C and heated again to the desired temperature. All reported transitions were obtained from the second heating portion of the cycle. Surface tension measurement was conducted using a dynamic contact angle (DCA) instrument (FTA200, First Ten Angstroms, Inc.) with a pendant drop method. A 3 mL plastic syringe and a needle with I.D. 0.483 mm and O.D. 0.711 mm were used. The aqueous drops were recorded by video and averages from at least 6 readings were used for reported values.

SEC

The SEC system consisted of a Waters Corporation 2695 solvent delivery system, 2487 dual wavelength spectrophotometric detector, 410 differential refractive index (DRI) detector, an Agilent (formerly Precision Detectors) PD2020 two-angle light scattering (LS) detector, and a Malvern (formerly

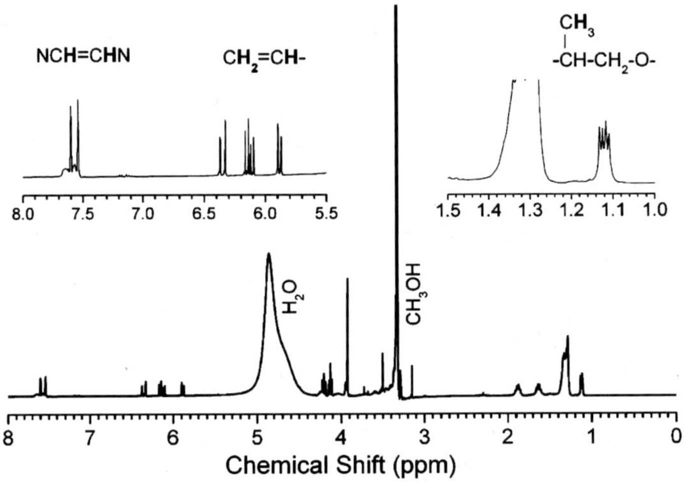

Figure 3. ^{1}H NMR spectrum of ATRP reaction mixture diluted in deuterated methanol. The left upper insert illustrates the detail for two of the imidazolium protons and for the vinyl protons. The right upper insert illustrates the detail for the propylene oxide methyl protons

Viscotek) H502 differential viscometry (DV) detector. The DV and DRI were configured with a parallel split after the spectrophotometric and LS detectors. The column set consisted of three 8 mm × 300 mm KF-806L columns from Shodex, thermostated at 40°C, calibrated with narrow-molecular-weight distribution poly(methyl methacrylate) (PMMA) standards. The eluent was 1,1,1,3,3,3-hexafluoro-2-propanol (HFIP) containing 0.01M tetraethylammonium nitrate, delivered at a nominal flow rate of 1.0 mL/min. The actual flow rate was determined from the retention volume of acetone added to the sample solvent as a flow marker.

Syntheses of Macroinitiators

The procedures are similar to those reported in the literature with modification (*7, 8*). For the preparation of macroinitiator **I** (Fig. 1) poly(propylene glycol) with a M*n* ~ 1000 (17 repeating units) and hydroxyl value of 111 mg KOH/g (8.0 g, 8 mmol) was dissolved in 120 mL anhydrous toluene. After azeotropic distillation of ~20 mL of toluene by rotary evaporator to remove traces of water, triethylamine (2.43 g, 24 mmol) was added, and the mixture was cooled to 0°C in an ice-water bath. In an argon atmosphere, 2-bromoisobutyryl bromide (BiBB) (5.52 g, 24 mmol) in 30 mL toluene was added dropwise over 50 min to the reactor. The reaction mixture was stirred for 24 hours at room temperature. After filtration, the filtrates were evaporated to remove most of the solvent (~20 mL left) on a rotary evaporator. The residue was dissolved in 120 mL methylene chloride and extracted with saturated NaHCO3 solution (3×50 mL). The organic phase was then dried over anhydrous NaSO4 and the solvent was removed using a rotary evaporator. The product was further dried in vacuo at 90°C for 3 hours to give a brown viscous liquid abbreviated as Br- PPG17-Br (**I**). A yield of 10.4 g (100%)

Table 1. Summary of molecular weight data for the poly(ILBr) based triblock copolymers synthesized by ATRP[a]

Sample	*Dp (NMR)*	*p (NMR)*	*M_n (NMR)*	*M_n (SEC)*	*M_w (SEC)*
TBI	10	96.3 %	4,800	7,560	58,200
TBII	20	74.4 %	12,300	17,800	56,000

[a] SEC of PILBr-*b*-PPG-*b*-PILBr in 1,1,1,3,3,3-hexafluoro-2-propanol (HFIP), light scattering detection. Degree of polymerization from ^{1}H NMR in CD_3OD

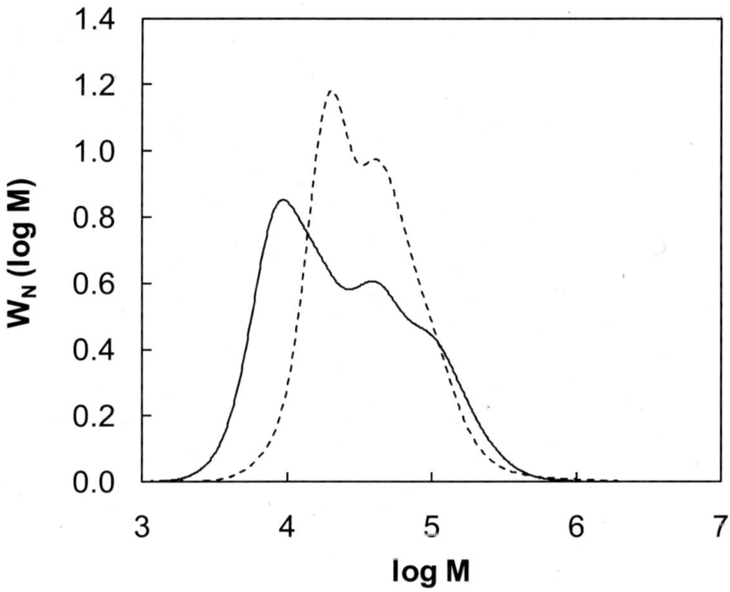

*Figure 4. Molecular weight distributions of triblocks **TBI** (solid, trimodal distribution) and **TBII** (dashed, bimodal distribution). These data are obtained using PMMA molecular weight standards and this distribution is a PMMA-equivalent molecular weight distribution.*

was obtained. ^{1}H NMR indicated that the degree of esterification was complete. Macroinitiator **II**, Br-PPG60-Br, was prepared following a similar procedure as above from a poly(propylene glycol) with a M*n* ~ 3500 (60 repeating units).

Synthesis of PILBr-*b*-PPG-*b*-PILBr Triblock Copolymers via ATRP of Difunctional Macroinitiators

Macroinitiator was added to a flask with argon pre-saturated methanol and water (maintained at a volume ratio of 13:8.5), followed by the addition of Cu(I)Br. This solution was bubbled with argon for 5 minutes. ILBr monomer was then added and bubbling was continued for 8 minutes. Ligand (HMTETA) was last added to the flask by micropipette followed by an additional 3 minutes bubbling. The reaction flask was installed with a condenser connected with an

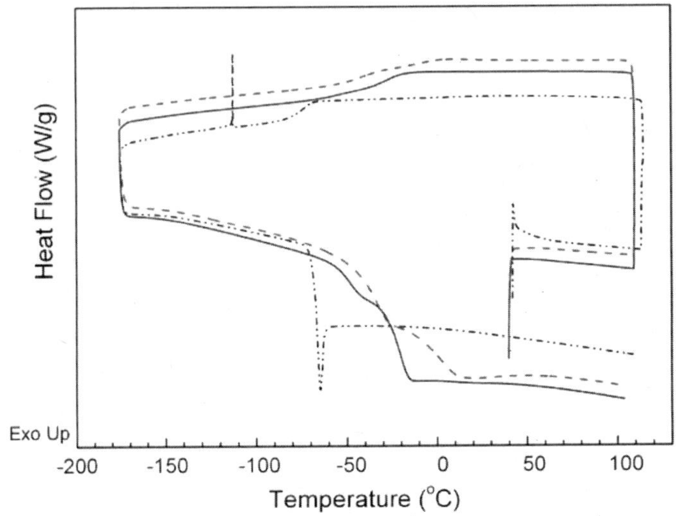

*Figure 5. DSC traces: – · · – Br-PO$_{17}$-Br (macroinitiator **I**), – – – poly(ILBr) (6), and — (IL)$_x$-(PO)$_{17}$-(IL)$_x$ (**TBI**).*

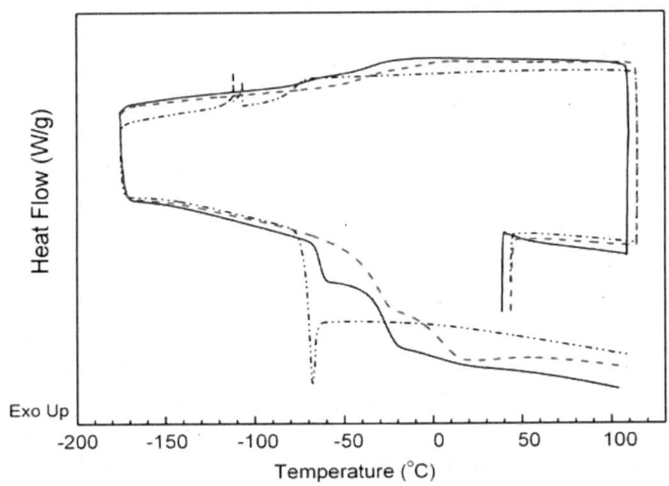

*Figure 6. DSC traces: – · · – Br-PO$_{60}$-Br (macroinitiator **II**), – – – poly(ILBr) (6), and — (IL)$_y$-(PO)$_{60}$-(IL)$_y$ (**TBII**).*

argon balloon and a vacuum inlet. Vacuum-argon cycling was performed for 10 minutes and the flask was immersed in a preheated oil bath at 40°C. The scheme for this reaction is illustrated in Fig. 2. Samples were periodically withdrawn using a nitrogen purge gas-tight syringe. The samples were diluted with CD_3OD for NMR analysis. The polymerization was terminated by purging air into the reaction. After polymerization, the solution was dialyzed against water in SnakeSkin® Pleated Dialysis Tubing (MWCO 3500). Dialysis was done for two days with water changes 3 times a day. Purified polymers were

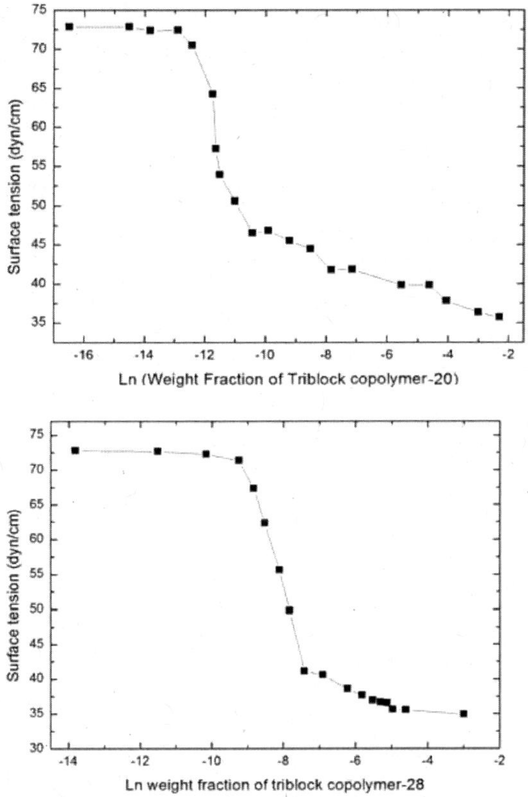

*Figure 7. Surface tension at the air-water interface for two triblock copolymers, **TBI** (top) and **TBII** (bottom) at 25°C.*

recovered by lyophilization and dried in vacuo. Macroinitiators **I** and **II** were used, respectively, to prepare (triblock) **TBI** and **TBII**.

Characterization

Conversion

The conversion of ILBr was estimated from ^{1}H NMR spectra, such as shown in Fig. 3. The imidazolium proton resonance relative area for the =N-CH=CH-N= protons over 7.40-7.80 ppm, divided by two, represents the sum of unreacted and polymerized monomer. The vinyl proton H_2C=CH- area over 5.7-6.5 ppm, divided by three, represents the unreacted ILBr monomer. The quotient of these quantities gives 1 minus the fractional conversion (p) of ILBr. While these initial analyses looked quite good, extensive SEC analysis showed that the preparation was much less controlled than we had hoped.

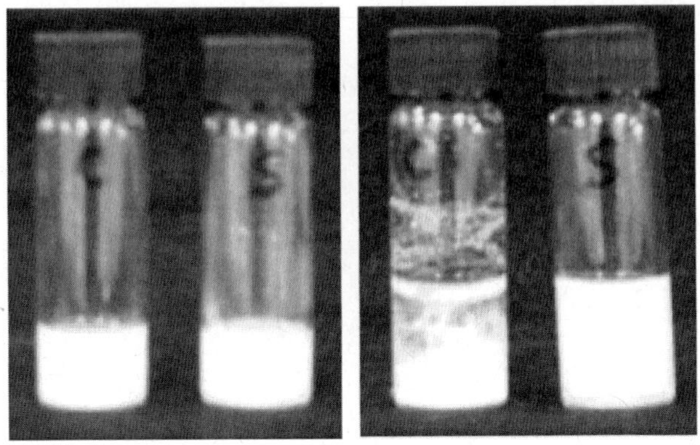

*Figure 8. Stabilization of polystyrene-sulfate beads (110 nm) using 0.101% by weight **TBII** solution. (Left) The vial marked C on the left contains 0.2 mL of DI water and 1.0 mL of 0.15 w/v Invitrogen beads. The vial marked S contains 1.0 mL of 0.15 w/v Invitrogen beads and 0.2 ml of **TBII** (0.00101 g/ml). (Right) On the right we see the same vials as on the left, except after addition of 1.0 ml of 1.0 M aqueous KBr. The illustrated sedimentation is after 16 h.*

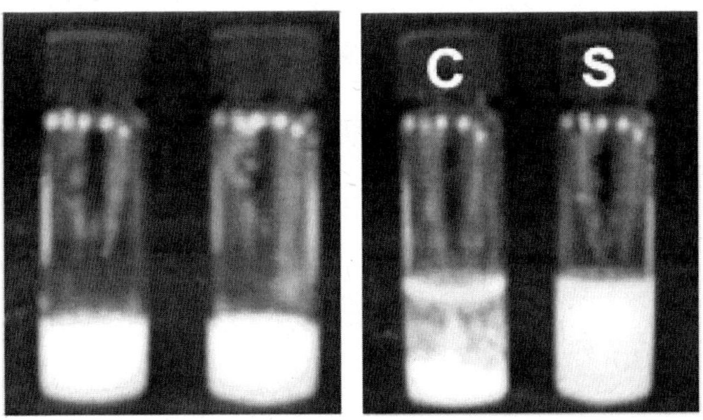

*Figure 9. Stabilization of polystyrene-sulfate beads (500 nm) using 1.01% by weight **TBII** solution. (Left) The vial marked C on the left contains 0.2 mL of DI water and 1.0 mL of 0.15 w/v Invitrogen beads. The vial marked S contains 1.0 mL of 0.15 w/v Invitrogen beads and 0.2 ml of **TBII** (0.0101 g/ml). (Right) On the right we see the same vials as on the left, except after addition of 1.0 ml of 1.0 M aqueous KBr. The illustrated sedimentation is after 16 h.*

Molecular Weight Distribution

The measurement of molecular weight distributions of block copolymers using light scattering and differential viscometry detection is complicated by the possibility of variation in the specific refractive index increment, *dn/dc*, across the chromatogram. This introduces errors in the calculation of concentrations from the DRI detector response at each retention volume when applying the typical assumption of constant *dn/dc* across the chromatogram. When combined with the

LS or DV signals, the uncertainty in concentrations can introduce inaccuracies in the molecular weights at each retention volume and distortions to resulting molecular weight distributions. In such instances, the shapes of the molecular weight distributions are better compared as simple PMMA equivalents using only the DRI detector signal and a narrow standard calibration curve. In the method, the local concentration error is not propagated by combining it with light scattering or viscometry detector signals. The shapes of the distributions may be closer to true although PMMA-equivalent molecular weight averages may not be correct. Figure 4 shows the differential weight fraction PMMA-equivalent molecular weight distributions obtained by our SEC analyses of **TBI** and **TBII**. Both products are much broader in molecular weight than expected for ATRP and subsequent analysis showed this defect was attributable to over saturation with macroinitiator at the beginning of the reaction. **TBI** appears broader than does **TBII**, and the **TBI** distribution is trimodal, with PMMA-equivalent modes at 9,260, 39,600, and 102,000 Da and PDI = 3.31. The **TBII** distribution appears to be bimodal, with modes at 20,000 and 41,300 Da and PDI = 2.04. Absolute number average and weight average molecular weights estimated from light scattering detection using measured values of *dn/dc* = 0.200 and 0.202 mL/g for TBI and TBII, respectively, are provided in Table 1 with conversion data from the NMR analyses. The SEC number average molecular weights are reasonable agreement with NMR values given their uncertainty because of the variable *dn/dc* problem.

The ILBr monomer absorbs at 270 nm. SEC-UV detection showed that the proportion of ILBr monomer in both TBI and TBII increased with increasing logM. This is consistent with a fixed middle block of PO with variable length end blocks of ILBr, with end block length increasing with overall logM.

Differential Scanning Calorimetry

The DSC for the macroinitiators **I** and **II**, triblock copolymers **TBI** and **TBII**, and poly(ILBr)$_x$ homopolymer produced (*6*) by ATRP are shown in Fig. 5 and Fig. 6, respectively. The distinct T_g evident in macroinitiator **I** at -65°C appears suppressed in **TBI**, while it appears present but shifted to 60°C in **TBII**. The double T_g behavior of the poly(ILBr) oligomer appears evident in both triblocks to some extent.

Surface Activity at the Air-Water Interface

Figure 7 illustrates surface tension at the air/water interface as a function of ln(weight fraction) for triblocks **TBI** and **TBII**. **TBI** actually has a M_w very close to that for **TBII**, and an apparently lower cmc (ln cmc_{TBI} = -10.5; cmc = 2.8 x 10^{-5} (weight fraction); ln cmc_{TBII} = -7.6; cmc = 5.0 x 10^{-4} (weight fraction). Both triblocks exhibit further surface tension lowering upon dissolving above the respective cmc and ultimately yield surface tensions at 25°C of 35 dyn/cm. The lower cmc exhibited by **TBI** is attributable to its broad molecular weight distribution (Fig. 4). The smallest molecular weight components of **TBI** correspond to PPG central blocks having end blocks of one to a few ILBr units, and these species are expected to be very surface active at low concentration. In addition, forming hairpin loops of the PPG central block is expected to be easier for the 17 PO block in **TBI** than for the 60 PO blocks of **TBII**.

Both of these triblocks are more surface active than the polyILBr homopolymer (*6*). The homopolymer surface tension behavior appears similar to that for **TBII** in Fig. 7(bottom) as the turn down in surface tension occurs at about the same weight fraction, and values of surface tension of 60, 50, and 45 dyn/cm are obtained at ln(weight fraction) of -8, -6, and -4, respectfully.

Immunization against Debye-Hückel Screening Induced Coagulation

The ultrastability observed (*1–3*) for nanolatexes derived from ILBr suggests that triblock copolymers having poly(ILBr) blocks may not be sensitive to indifferent electrolytes, but only to specific stimuli inducing anions. To test this hypothesis we prepared suspensions of highly negatively charged polystyrene beads, 110 and 500 nm in diameter, and restabilized some of them by adsorbing what was estimated as an excess of **TBII**. These preparations are illustrated on the left sides of Figs. 8 and 9 wherein the control (C) is stabilized with surface sulfate groups and the samples (S) have been restabilized with adsorbed **TBII**.

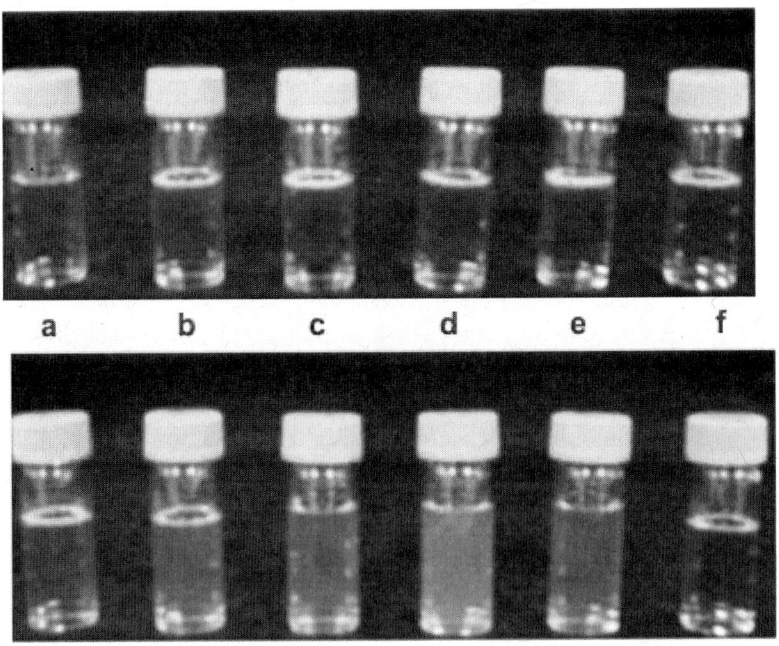

*Figure 10. The upper series of vials each contain 2 ml DI water and 100 μl of aqueous 1.5% (w/v) of **TBII**. The lower series of vials are the same as the upper series, but after addition of 0.01 M KPF_6 solution in the amounts of 50 μl (a); 100 μl (b); 150 μl (c); 200 μl (d); 250 μl (e); and none (f).*

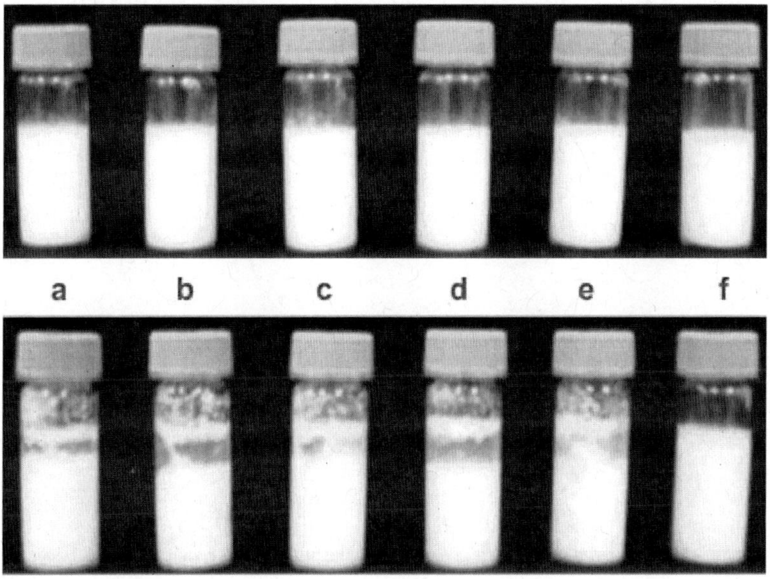

*Figure 11. The upper series of vials each contain 2 ml of 4.0% (v/v) UCAR 300. Vials (a) through (e) had 75 μl of 1.5% (w/w) aqueous **TBII** added (none added to vial f). The lower series of vials are the same as above but after addition of 0.01 M aqueous KPF_6 in the amount of 100 μl (a); 150 μl (b); 200 μl (c); 250 μl (d); 300 μl (e); 250 μl (f).*

When concentrated aqueous KBr is added to each vial, two things happen. In the cases of the sulfate stabilized beads in the "C" vials, the added aqueous KBr causes rapid coagulation due to the electrostatic screening provided by the high concentration of indifferent electrolyte, KBr. In the vials wherein the beads have been restabilized with excess **TBII**, the beads are immunized against this high electrolyte induced coagulation, and the bead suspension remains unperturbed and stable. These respective coagulation and continued stable suspension effects are clearly seen for both bead sizes in the right-hand photos of Figs. 8 and 9.

Stimuli-Responsive Behaviors

Figure 10 illustrates a selective precipitation behavior exhibited by a **TBII** solution when varying amounts of KPF_6 are added. The hexafluorophosphate anion interacts strongly with imidazolium rings and transforms these rings from hygroscopic moieties into hydrophobic centers due to the formation of a strong ion pair. This decrease in water solubility results in the formation of insoluble particles, shown by dynamic light scattering to be of the order of 200-300 nm in diameter. This particle formation is indicated by the turbidity seen in the lower series of vials a-e in Fig. 10.

Another interesting behavior induced by the addition of the hexafluorophosphate anion is illustrated in Fig. 11. Here a commercial latex dispersion normally not sensitive to indifferent or other salt levels was sensitized

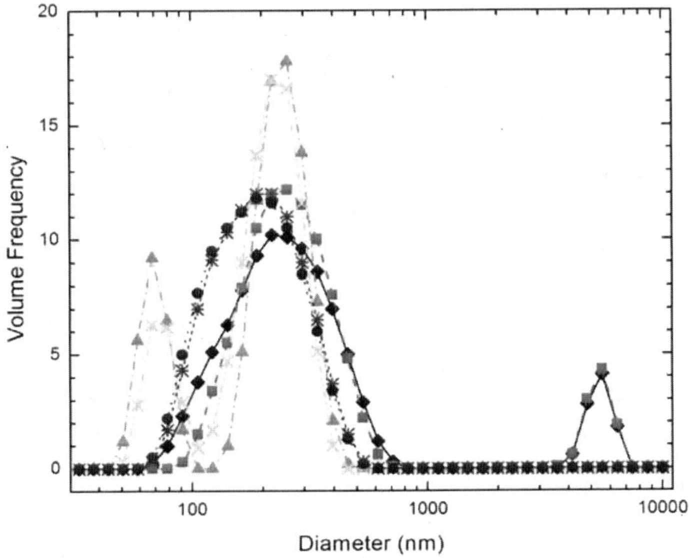

*Figure 12. Particle size distributions of PY138/**TBII** aqueous dispersions determined by dynamic light scattering. The yellow (–▲) and turquoise (–×–) plots (800 nm filter using high resolution data conversion) show that the 200-300 particle size is gradually being reduced to about 60 nm. The magenta (–■–) and blue (–◆–) plots (no filtration with normal settings) show there is a bit of loose aggregate in the multimicron size regime. The reddish-brown (–●–) and purple (–*–) curves (800 nm filter using normal settings) shows that the distribution is broad, going from about 80 to about 300 nm at peak and further.*

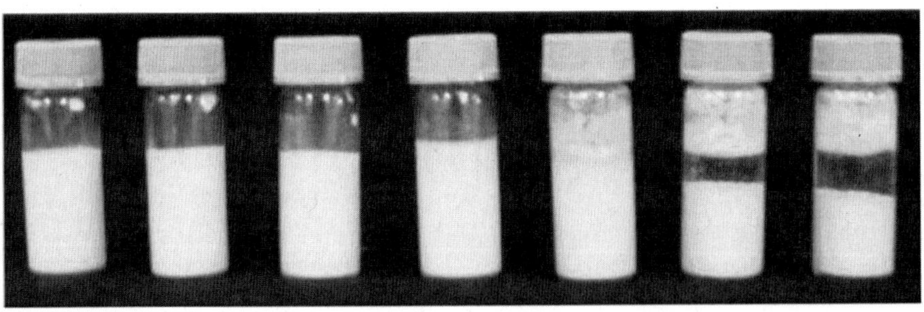

*Figure 13. Stimuli responsiveness of the PY138/**TBII** dispersions after addition of 0.1 M aqueous KPF_6 to equal aliquots of the dispersion. The vial at far left had only DI water added as a diluent. Then increasing amounts of KPF_6 were added with shaking. The significant sedimentation illustrated in the right most two vials shows that these dispersions were significantly destabilized by the added KPF_6. The dispersion at the lower level used in the vial third from right was also partially destabilized, as it is plain to see that some of the dispersion particles therein stick to rather than drain away from the inner glass wall.*

by adsorbing **TBII** from solution. The vials (a) through (e) in Fig. 10 had varying amounts of destabilizing KPF_6 added, and the resulting coagulation illustrated the effect, induced by the sequential adsorption of **TBII** followed by addition of the hexafluorphosphate anion. The vial (f) had no **TBII** added while it did have an appreciable amount of KPF_6 added, and illustrates the basic insensitivity of the latex to the added salt in the absence of the stimuli responsive triblock.

In Fig. 12 we illustrate particle size (volume) distributions obtained by jar milling, overnight, some pigment yellow 138 (PY138) in water using 100 μm ceramic beads and **TBII** to stabilize the dispersion. Approximately 30 mL of dispersions was milled that comprised about 1.2g PY138 pigment and about 100 mg of **TBII** as dispersant. After milling the resulting dispersion was separated from the milling media using a fine weave filtration bag.

The highest resolution data of Fig. 12 show that the 200-300 nm particles are being reduced further to about 60 nm, and illustrate the high efficacy of the **TBPII** as an aqueous dispersing aid. The effects of filtration show that there is some weak aggregation of particles with clusters between 1 and 10 μm (Fig. 12). We expect that sub-100 nm size would be obtainable with longer milling time. However, this degree of dispersion suffices for the ensuing stimuli responsive experiment.

We then took equal aliquots of this dispersion and added varying amounts of KPF_6 to destabilize the dispersion, relying on the stimuli responsiveness of the ILBr blocks in the **TBII**. The photos in Fig. 13 illustrate this destabilization as evidenced by the sedimentation shown as well as by the stickiness of some of the dispersions (third vial from the right in Fig. 13) to the interior vial surface. In the absence of sufficient destabilizing KPF_6, the resulting pigment dispersion is very stable, as demonstrated by the left most four vials in Fig 13.

The left most vial in Fig. 13 only had water added rather than any KPF_6 solution. The effects of added KPF_6 on the right most three vials in Fig. 13 were immediate, but the degree of sedimentation gradually increased with time.

Conclusions

ATRP (but without good polydispersity control) was used to create a new class of triblock copolymers somewhat analogous to the Pluronic® series. Our **TBI** and **TBII** examples have center blocks of about 16 (1,000 Da) and 60 (3,500 Da) PO groups, respectively. The endblocks ranging from 1 to more than 100 ILBr groups impart additional surface activity, because of the undecyl groups, and aqueous solubility from the hygroscopic imidazolium bromide groups.

Very effective surface activity of these new triblocks was demonstrated at the air/water interface. Introductory milling experiments also showed that this surface activity extends to the pigment/water interface.

A remarkable feature of these new triblocks is that the ILBr blocks appear to behave as effectively nonionic. We showed for the case of two polystyrene charged beads that restabilization with **TBII** immunizes the beads from salt induced coagulation (from Debye-Hückel).

On the other hand, addition of a stimuli responsive anion, such as PF_6^-, induces stimuli responsiveness. Stimuli responsive precipitation and dispersion destabilization were demonstrated.

More recent synthetic refinements have shown that excellent control is achievable (PDI ~ 1.2–1.7) and that some such triblocks are distimuli responsive exhibiting thermoreversibility in addition to anion responsiveness.

References

1. Yan, F.; Texter, J. *Chem. Comm.* **2006**, 2696–2698.
2. England, D. MS Thesis. Eastern Michigan University, 2008.
3. Tambe, N. MS Independent Study Report. Eastern Michigan University, 2009.
4. Tauer, K.; Weber, N.; Texter, J. *Chem. Comm.* **2009**, 6065–6067, doi: 10.1039/B912148J.
5. Alexandridis, P.; Holzwarth, J. F.; Hatton, T. A. *Macromolecules* **1994**, *27*, 2414–2425.
6. Ma, X.; Ashaduzzaman, Md.; Kunitake, M.; Crombez, R.; Texter, J.; Slater, L.; Mourey, T. *Langmuir* **2011**, in press.
7. Liu, S.; Armes, S. P. *J. Am. Chem. Soc.* **2001**, *123* (40), 9910 9911.
8. Li, C. H.; Ge, Z. S.; Jin, F.; Liu, S. Y. *Macromolecules* **2009**, *42* (8), 2916–2924.

Editors' Biographies

Patrick Theato

Patrick Theato was born in Wiesbaden, Germany in 1973. He studied chemistry at the University of Mainz and the University of Massachusetts, Amherst, and obtained his Ph.D. degree under the supervision of Prof. R. Zentel at the University of Mainz in 2001. Shortly after, he was awarded a Feodor Lynen Postdoctoral Research Fellowship from the Humboldt Foundation and joined the group of Prof. D.Y. Yoon at Seoul National University (Korea), where he worked as a postdoctoral fellow, followed by a short research stay at Stanford University with Prof C.W. Frank. In 2003, he joined the University of Mainz as a young faculty and completed his Habilitation in 2007. Since 2009 he holds a joint appointment with the School of Chemical and Biological Engineering at Seoul National University within the WCU program. In 2011 he accepted a prize senior lectureship at the University of Sheffield, UK. Shortly after he moved to University of Hamburg, Germany, where he is currently a professor for polymer chemistry. His current research interests include the defined synthesis of reactive polymers, design of multi stimuli-responsive polymers, versatile functionalization of interfaces, hybrid polymers, polymers for electronics and templating of polymers.

Andreas F. M. Kilbinger

Andreas Kilbinger was born in Mainz, Germany in 1971. He studied chemistry at the University of Mainz and the Free University of Berlin and received his PhD degree from the University of Durham (United Kingdom), working in the group of W.J. Feast. With a Feodor-Lynen Fellowship awarded by the Alexander von Humboldt Foundation he carried out research as a postdoctoral scholar in the group of R.H. Grubbs at the California Institute of Technology (Pasadena, CA, USA). After that he returned to the University of Mainz as a junior faculty member and received his Habilitation in 2007. In the same year he was awarded the Herrmann Schnell-Award by the German Chemical Society (GDCh).

Since 2010 he is working as a professor for polymer chemistry at the Université de Fribourg, Switzerland. He is interested in the synthesis of precision polymers, including the development of new polymer synthetic methods for introducing functionality and all aspects of monomer sequence control in synthetic polymers. The research aims at finding synthetic methods that bridge the divide between nature's precisely defined macromolecules and today's synthetic polymers.

E. Bryan Coughlin

E. Bryan Coughlin, PhD is a Professor at The University of Massachusetts Amherst in the Department of Polymer Science and Engineering. Dr. Coughlin is actively engaged in the development of novel polymeric materials for advanced applications in aerospace, sporting goods, electronics, packaging, and numerous other industrial sectors that utilize plastics. Upon the completion of his Doctorate in Chemistry at the California Institute of Technology, he joined the Central Research and Development Department of the DuPont Company where he was on staff for nearly 6 years. Dr. Coughlin is a co-inventor of the DuPont's Versipol® Polyolefin Technology Platform, and has over 25 patents to his name. His expertise covers anionic and radical polymerization as well as the use of organo-transition metal catalysts for the preparation of well-defined polymers. Since 1999 Dr. Coughlin has been on the Faculty of the Polymer Science and Engineering Department at UMass Amherst. He has won a number of national research awards.

Indexes

Author Index

Subject Index

Q

R

S

T

V